First European Conference on

INTEGRATED OPTICS

14 - 15 September 1981

Organised by the

Electronics Division of the Institution of Electrical Engineers

in association with the

Institute of Mathematics and its Applications
Institute of Physics

with the support of the

Convention of National Societies of Electrical Engineers of Western Europe (EUREL)

Venue

The Institution of Electrical Engineers, Savoy Place, London WC2, UK

Programme Committee

J E Midwinter *(Chairman)*
British Telecom Research Laboratories, UK

F Auracher
Siemens AG, Federal Republic of Germany

J Buus
Technical University of Denmark, Denmark

P J R Laybourn
University of Glasgow, UK

D B Ostrowsky
University of Nice, France

M Papuchon
Thomson-CSF, France

S D Smith
Heriot-Watt University, UK

W Sohler
Fraunhofer Institut für Physikalische Messtechnik,
Federal Republic of Germany

S Sottini
Istituto di Ricerca sulle Onde Elettromagnetiche, CNR, Italy

W J Stewart
Plessey Research (Caswell) Ltd, UK

R Ulrich
Technische Universität Hamburg-Harburg,
Federal Republic of Germany

Local Organising Committee

J E Midwinter *(Chairman)*
British Telecom Research Laboratories, UK

I Childs
British Broadcasting Corporation, UK

A W Horsley
Standard Telecommunication Laboratories Ltd, UK

P J R Laybourn
University of Glasgow, UK

S D Smith
Heriot-Watt University, UK

W J Stewart
Plessey Research (Caswell) Ltd, UK

B A Tozer
Central Electricity Generating Board, UK

Published by the Institution of Electrical Engineers, London and New York, ISBN 0 85296246 0. ISSN 0537-9989.

The cover was designed by Thomas/Weintroub Associates.

Produced by Thomas/Weintroub Associates, London.

© 1981 The Institution of Electrical Engineers.

Contents

Contents

Contents

Contents

List of Authors

DIFFUSED OPTICAL WAVEGUIDES ON THE +c AND -c FACE OF c-PLATE LiNbO$_3$ - CHARACTERISTICS REVEALED BY CHEMICAL ETCHING

D. Hoffmann and U. Langmann

Ruhr-Universität Bochum, Institut für Elektronik, W.-Germany

[handwritten note: 1-11-88 2:02 AM "We will make "hey" while the Son shines. That means that we will tell others about Jesus. WHH"]

INTRODUCTION

LiNbO$_3$ ranges among the most extensively used materials for integrated optics. This is due to the facts that it possesses strong electro-optic coefficients and that waveguide forma-tion can be realized by metal (e.g. Titanium) indiffusion into the ferroelectric crystal or or by simple Li$_2$O outdiffusion.

A crystal orientation widely used for inte-grated optical directional couplers is the z-cut (c-plate) orientation, with the optical crystal axis perpendicular to the substrate surface. Using this crystal orientation, effi-cient and fast directional couplers can be fa-bricated, if the metal electrodes cover the optical waveguides.

An important question, scarcely discussed in the literature, concerns the choice of the +c or -c face of the substrate for device imple-mentation. Miyazawa (1) reports a ferroelec-tric domain inversion due to Ti-indiffusion into the +c face. Furthermore, he reports a multidomain structure within Ti-diffused stripes. This result, if assumed to be correct, is of great importance for the device implemen-tation in c-plate LiNbO$_3$.

This contribution aims at revealing character-istic differences between optical waveguides on +c and -c faces of LiNbO$_3$ by chemical etch-ing. Furthermore, it generally wants to suggest a deliberate choice of the +c or -c face for LiNbO$_3$ integrated optical devices.

WAVEGUIDE IMPLEMENTATION AND ETCHING TECHNIQUE

Directional coupler structures (3 and 5 μm wide waveguides, 3 and 5 μm spacing between waveguides before diffusion) were fabricated on both the +c and -c face of c-plate LiNbO$_3$ (single-domain material of optical quality from Union Carbide). +c and -c faces were iden-tified by measuring the polarity of the pyro-electric voltage. The Ti layer thickness ranged from about 350 Å to 550 Å. The diffu-sion was performed under streaming air at 1010 °C for 5 hours. The heating up rate was chosen to be very low (about 100 °C/h up to 600 °C), in order to avoid waveguide destruction by py-roelectric influences. Li$_2$O-outdiffusion was not compensated for. After Ti-indiffusion, the waveguide relief exhibited straight edges. The samples were etched in 1 HF + 2 HNO$_3$ for 2 to 14 minutes at 100°C. This etch is known to be sensitive to the ferroelectric domain polarity.

Undiffused LiNbO$_3$ remains nearly unchanged on the +c face after etching, whereas the -c face shows etch hillocks of triangular or conical shape, as observed by Ohnishi and Iizuka (2).

WAVEGUIDE CHARACTERISTICS AFTER ETCHING

Ti-diffused Waveguides

The etching of Ti-diffused optical waveguides on both faces shows characteristic differences.

Stripe waveguides on +c faces.
Waveguides on the +c face change into grooves. The edges of the groove are formed by steep slopes. The bottom of the groove has a rough and scaly surface, which is evened up after further etching. The width of the etched waveguide pattern depends on the initial Ti concentration and, to some extent, on the time of etching. The width of the grooves is much larger than the original Ti stripe width. The depth and the depth profile of the grooves on +c faces are a function of the time of etching; we found a maximum etch rate of about 0.5 μm/min for an initial Ti layer of 350 to 550 Å on the +c face.

Fig.1 and 2 show two examples after etching for 6 minutes. Fig.1 is a photomicrograph of a groove which results from the coupling sec-tion of a directional coupler. Fig.2 shows a part of the diverging waveguides. The micro-scope is focussed onto the bottom of the grooves, which leads to the surrounding sur-face being out of focus. The depth of the grooves is about 3 μm.

High Ti concentrations (550 Å) cause trian-gular spots (hillocks) to appear as additio-nal etch patterns along the bottom of the grooves, as shown on the micrograph of Fig.3. Low Ti concentrations in the solid after dif-fusion, resulting e.g. from small stripes, thin Ti layers, and/or particular diffusion parameters, lead to partial etching on +c faces. Fig.4 is a SEM micrograph of etched waveguides (5 μm wide Ti stripes before dif-fusion, about 340 Å Ti layer thickness, pre-diffusion period of several hours at tem-peratures increasing from 850 to 1010 °C). This type of partial etching was observed by Miyazawa (1), as well.

Stripe waveguides on -c faces. Waveguides on the -c face lead to shallow grooves, as well. Their depth, however, is much smaller (less than 1 μm after etching for 9 minutes). The etch rate on the Ti-diffused -c face is, apparently, considerably less than on the +c face. Fig.5 and Fig. 6 show again a part of the coupling section and the diverging wave-guides, respectively. Steep slopes do not exist. Both the surrounding surface and the grooves are focussed, which demonstrates the little depth of the etched structures.

Outdiffused planar waveguides

After the diffusion process, the substrate surface without Titanium is generally supposed to become an outdiffused optical planar wave-guide by the outdiffusion if Li$_2$O. Etching of a +c face produces etch patterns as discernible on the SEM micrograph of Fig.4 . -c faces, however, do not reveal characteristic etch patterns after diffusion; they rather appear to remain unchanged by the diffusion process and show the etch pattern of an undiffused -c face.

TENTATIVE INTERPRETATION AND CONCLUSION

Etch patterns of Ti-diffused optical waveguides show a clear dependence on the Ti concentration. Apparently, the width of the etched stripes reflects the lateral Ti-diffusion along the whole waveguide structure. A quantitative interpretation of the diffusion profile, however, necessitates a detailed knowledge of the functional dependence of the etch rate on the Ti concentration in Ti-doped $LiNbO_3$. These data are not yet available. The characteristic difference between the etch rates on the positive and the negative surface of Ti-diffused c-plate material is not understood, to date.

The etching mechanism in the Ti-doped solid appears to be more complex than in the undoped material. Hence, the interpretation of the etch patterns under doping conditions should not be based on an inconsiderate application of results obtained with undoped crystals. The result of Fig.4, however, suggests that Ti-indiffusion on positive faces leads to a (partial) polarity reversal within the stripe waveguide. (This statement presumes that, even after Ti-indiffusion, the etch distinguishes between the polarization directions.) According to (1) and (2), the triangular spots (hillocks) of Fig.3 may, tentatively, be interpreted as microdomains of antiparallel orientation as compared to the surroundings.

The particular differences between non-Ti-diffused areas after etching suggest to examine carefully the outdiffusion behaviour on both faces. From an inspection of the crystal structure, it becomes apparent that Li ions have to transverse the oxygen layers in the process of polarity reversal (3). Therefore, the position of the Li ion relative to the oxygen layer characterizes the domain polarity. Hence, differences between the Li_2O-outdiffusion behaviour on +c and -c faces are probable.

In conclusion, the etching of Ti-diffused waveguides on negative faces yields less complex results, according to our experiments. This observation supports the suggestion of (1) to prefer -c faces for integrated optical device implementation on c-plate $LiNbO_3$. However, this suggestion is to be considered as tentative until it is supported by an additional comparison of the <u>optical</u> properties of Ti-diffused waveguides on the positive and negative face of c-plate $LiNbO_3$.

ACKNOWLEDGEMENT

We gratefully acknowledge the encouragement by Prof. Bosch and the financial support of the German Research Council (DFG).

REFERENCES

1. Miyazawa, S., 1979, <u>J. Appl. Phys.</u>, <u>50</u>, 4599 - 4603.
2. Ohnishi, N., and Iizuka, T., 1975, <u>J. Appl. Phys.</u>, <u>46</u>, 1063 - 1067.
3. Niizeki, N., et al, 1967, <u>Jap. J. Appl. Phys.</u>, <u>6</u>, 318 - 327.

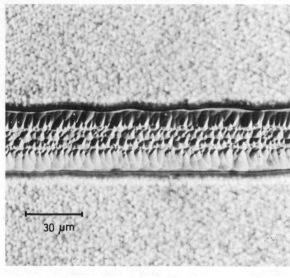

+c race

Directional coupler			Time of etching
width of waveguides before diffusion	5 µm		6 min
spacing of waveguides	5 µm		
Ti-layer thickness	350 Å		
diffusion at 1010 °C for 5 h			

Figure 1 Part of the coupling section: Enlarged micrograph of the etched groove (depth 3 µm).

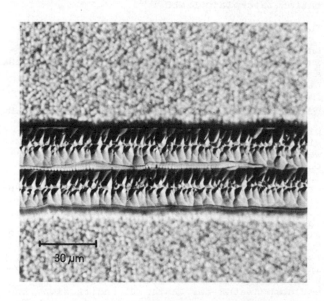

+c race

(See Fig.1 for the list of parameters)

Figure 2 Section of the diverging waveguides: Micrograph of etched grooves.

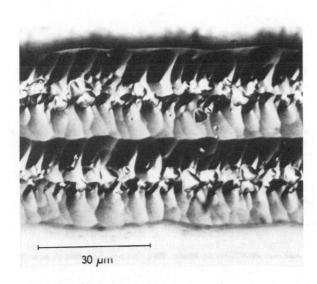

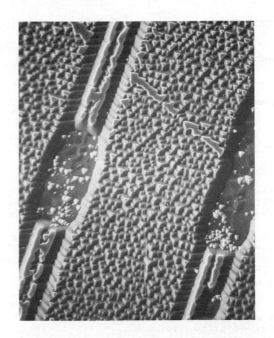

<u>+c face</u>

Ti-layer thickness   550 Å

Figure 3   Enlarged section of the diverging
waveguides: Etched grooves with
triangular spots (hillocks)

<u>+c face</u>

Ti-layer thickness 340 Å   Time of
                                            <u>etching</u>
Long prediffusion period
(cf. text)   5 min

Figure 4   SEM micrograph of waveguides
showing partial etching.

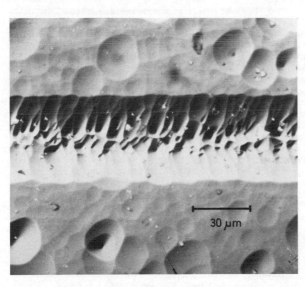

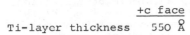

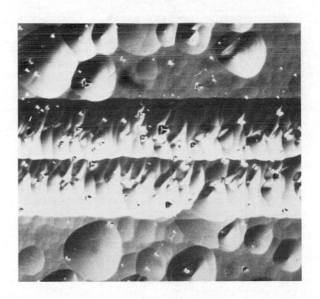

<u>-c face</u>

<u>Directional coupler</u>   Time of
                                            <u>etching</u>
  width of waveguides
    before diffusion   3 μm   9 min
  spacing of waveguides   3 μm
  Ti-layer thickness   460 Å
  diffusion at 1010 °C
          for   5 h

Figure 5   Part of the coupling section:
Micrograph of etched patterns.

<u>-c face</u>

(See Fig.5 for the list of parameters)

Figure 6   Part of the diverging waveguide
after etching.

TITANIUM DIFFUSED LiNbO$_3$ OPTICAL WAVEGUIDES

A.D. McLachlan*, R.M. De La Rue, J.A.H. Wilkinson

University of Glasgow, U.K.

INTRODUCTION

Titanium diffused LiNbO$_3$ waveguides are a central feature of many of the integrated optic devices which have been investigated over the last few years. Many active devices such as switches, phase and amplitude modulators, directional couplers, variable wavelength filters, electro-optic and acousto-optic deflectors have been successfully demonstrated.

The titanium diffusion process into LiNbO$_3$ can lead to low loss waveguides whose effective index of propagation (n_{eff}) can be accurately measured by means of the prism coupling technique (1). The exact characterisation of the waveguide refractive index profile is problematic because of uncertainties in the mechanism of the refractive index change, the diffusion pathways, and material effects such as high temperature solid state reaction chemistry, ferroelectric domain rotation and outdiffusion of LiO$_2$ and O$_2$. The small diffusion depths (typically 2 - 4 μm) mean that very high resolution techniques are needed to measure the refractive index or diffusion profiles.

In this paper we present a comparison between theoretical and experimental dispersion curves for Ti-LiNbO$_3$ waveguides as a function of fabrication parameters such as initial Ti film thickness, diffusion time, waveguide width and optical wavelength. We also present the results of a study on the effect of photorefractive damage on the propagation in channel waveguides.

TITANIUM INDIFFUSION MECHANISMS AND REFRACTIVE INDEX CHANGE

The nature of the diffusion mechanism has been studied by Sugii et al (2) who suggested that the Ti ions enter the LiNbO$_3$ lattice at the Nb sites and that the consequent change in refractive index is partly due to the strain introduced into the crystal by the Ti impurities, which have a different ionic radius from the Nb ions, partly due to the decreased spontaneous polarisation of lattice (due to the Ti impurity) and partly due to contributions from the reduced ionic polarisability of this substituting ion. Another possible contributory factor was put forward by Esdaile (3) who showed from the LiNbO$_3$ polarisability studies of Krindach et al (4) and the crystallographic investigation of Megaw (5) that the refractive index change could be due to a change in ion polarisability and a crystal structure rearrangement involving a rotation of the oxygen octehedra to accommodate the diffusing ions.

Whatever the exact details of the mechanisms involved, it is important for integrated optic devices that Ti diffused LiNbO$_3$ waveguides be precisely characterised. Some previous papers have studied the diffusion

*Now with Racal-MESL Ltd, U.K.

and refractive index change by observation of the mode spacing in slab waveguides (6,7) or by investigation of the Ti concentration by electron microprobe analysis (2,7,8,9,10).

If the diffusion of a thin Ti film into a LiNbO$_3$ crystal could be approximated by the limited source simple diffusion case then the appropriate solution to the 1-dimensional diffusion equation would be a Gaussian diffusion profile. The experimental results (2,6,7, 8,9,10) showed that a Gaussian profile could be an adequate approximation to the refractive index profile but that effects such as LiO$_2$ outdiffusion and surface diffusion could introduce further discrepancies from the theoretically predicted simple profile. The finite resolution capability ($\sim$ 1 μm) of the electron microprobe method means that small ($\sim$300 A) surface or sub-surface layers cannot be accurately observed.

The use of Secondary Ion Mass Spectrometry (SIMS) has allowed a more detailed study of the Ti concentration as a function of depth into the substrate. Burns et al (7) observed a Gaussian depth diffusion profile with an additional peak, having a diffusion depth of the order of 0.3 μm, directly below the crystal surface. They postulate that this may be due to the formation of a thermodynamically preferred LiTiO$_3$ state. Further evidence of this can be seen in the evolution of the titanium layer as diffusion progresses (3,11). The titanium is oxidised, then starts to form an apparently multiphase polycrystalline layer which subsequently smooths out after prolonged diffusion. The thickness of this layer has been measured to be approximately 50% thicker than the original Ti film. It seems probable that this layer will have a refractive index value at least as high as the peak surface index of the LiNbO$_3$, due to the high refractive index of the constituent materials (for TiO$_2$: n_e = 2.9, n_o = 2.6). Both the possible formation of LiTiO$_3$ and the high index loading layer will introduce further discrepancies from results predicted by the simple Gaussian profile.

TI-DIFFUSED LiNbO$_3$ SLAB WAVEGUIDES

In order to model the dispersion of slab waveguides we have examined two possible models for the refractive index profile (Fig. 1). Fig. 1a shows the Gaussian solution to the diffusion equation while the profile of Fig. 1b shows a profile which approximates the titanium concentration profile found by the SIMS investigations of Burns et al (7). Both models have a surface layer, typically 300-1000 Å, which is 1.5 times the original Ti film thickness and in which the index is assumed to be the same as the peak substrate index. This represents the residual raised layer left by the oxidised Ti film when it diffuses into the bulk of the substrate.

Typical diffusion depths are 3.2 μm for the

Gaussian profile and 0.45 µm and 3.2 µm for the "double Gaussian" profile when using an 18 hour diffusion at 1000°C.

We have examined the propagation of TE modes in slab waveguides by the numerical analysis of a multilayer structure (12). The parameters of this model have been fixed by comparing experimental and theoretical results for the modal dispersion as a function of the initial titanium film thickness. With these model parameters fixed, the waveguide dispersion characteristics as a function of other fabrication parameters were calculated.

The slab waveguides were fabricated by the electron beam evaporation of titanium onto Y-cut LiNbO₃ substrates. In order to prevent out-diffusion of LiO_2 these waveguides were then diffused at 1000°C, under equilibrium conditions, in a sealed system (13).

The effective propagation constants (n_{eff}) of the TE (or E_z) modes were measured with a prism coupler. Measurement of the bulk LiNbO₃ index was also made by observing the cut-off angle for the substrate mode and the value of Δn_{eff}, i.e. ($n_{eff} - n_{substrate}$), was calculated. The corresponding theoretical curves were computed and compared with the experimental dispersion curves (Fig. 2).

Light propagation in several slab waveguides has been measured at a number of wavelengths in the range from 458 nm to 633 nm (Argon ion laser, Rhodamine 6G dye laser and HeNe laser) and the experimental results were compared with the corresponding theoretical curves (Fig. 3). For this theoretical model it is assumed that the fractional index difference due to titanium indiffusion is constant over the wavelengths involved. The bulk index for LiNbO₃ at the various wavelengths was calculated from the modified Sellmeir equations presented by Smith et al (14).

Results will also be presented comparing theoretical and experimental dispersion curves as a function of diffusion time, involving the modelling of oxidised Ti film as it diffuses into the crystal surface.

TI-DIFFUSED LiNbO₃ CHANNEL WAVEGUIDES

The theoretical diffusion profile of the slab waveguide can be extended to the channel waveguide situation by assuming separation of variables in the diffusion problem (15). This is permissible if, in uniaxial crystals, the diffusion axes coincide with the crystal axes. The channel waveguide Z-direction diffusion profile, perpendicular to the waveguide direction and parallel to the crystal surface, has been calculated by Hocker and Burns (16), and a complete expression for the refractive index profile of a channel waveguide, of width w, is given by:-

$$ n(y,z) = n_{sub} + \Delta n(0) \cdot \left[\exp\left[-\left[\frac{y}{D_y}\right]^2\right] \cdot \frac{1}{2} \cdot \left[erf\left[\frac{w+2z}{2D_z}\right] + erf\left[\frac{w-2z}{2D_z}\right] \right] \right] $$

The theoretical evaluation of these channel waveguides has been achieved by using a variational-type numerical approach to the wave equation for the guided TE modes (17).

These waveguides were investigated experimentally both by prism input/output coupling and end fire input/output coupling. Fig. 4 shows the plot of results obtained by end-fire coupling to several waveguides (of different widths) with varying initial

titanium film thicknesses. The experimental points are the number of modes observed and the theoretical curves are the calculated mode cut-off curves. The modal energy patterns of the guided modes can easily be calculated using the variational technique and Fig. 5 shows the mode patterns for one such channel waveguide.

The phenomenon of 'optical damage' has long been known to occur in LiNbO₃ crystals when they are illuminated with intense light at wavelengths in the visible part of the spectrum. Chen (18) observed a decrease in n_e and n_0 of LiNbO₃ when illuminated with red (0.6328 µm) and green (0.488 µm) light but that no damage occured at near-infrared wavelengths (1.064 µm). The exact nature of the photorefractive effect has been examined by Glass (19).

In this paper we present the results of an investigation of optical waveguide propagation at relatively high power densities at 0.512 µm. One aspect of this study was the variation with time in the power propagated by a 10 µm channel waveguide (400 Å Ti, 18 hours diffusion at 1000°C). Fig. 6 shows this variation with time as a function of initial waveguide throughput. It can be seen that the higher the throughput power the more pronounced the effect of photorefractive damage. A lower threshold power of 40 µW was observed: below this figure no time variation of the guided mode power was observed. This corresponds to a threshold power density of approximately 200 W.cm⁻². The explanation of this phenomenon appears to be that the refractive index of the waveguide is reduced due to the photorefractive effect and that this eventually leads to the higher order modes becoming less well guided and leaking out into the substrate, causing the observed loss of guided mode power.

ACKNOWLEDGEMENTS

The authors would like to thank C.D.W. Wilkinson and R.J. Esdaile for helpful discussions. A McLachlan was supported by a Science Research Council CASE award with the Plessey Co. Facilities for part of this work were made available by Mr. W. Stewart of the Plessey Allen Clark Research Centre and are gratefully acknowledged.

REFERENCES

1. Tien, P.K., and Ulrich, R., 1970, J. Opt. Soc. Am., 60, 1325-1337.

2. Sugii, K., Fukuma, M., and Iwasaki, H., 1978, J. Mat. Sci., 13, 523-533.

3. Esdaile, R.J., 1979, Ph.D. Thesis, University of Glasgow.

4. Krindach, D.P., Maiorow, W.S. and Meisner, L.B., 1976, Sov. Phys. Solid State, 18, 1756-1757.

5. Megaw, H.D., 1968, Acta. Cryst. A24, 583-588.

6. Naitoh, H., Nunoshita, M., and Nakayama, T., 1977, Appl. Opt., 16, 2546-2549.

7. Burns, W.K., Klein, P.H., and West, E.J., 1878, J. Appl. Phys., 50, 6175-6182.

8. Fukuma, M., Noda, J., and Iwasaki, H., 1978, J. Appl. Phys., 49, 3693-3698.

6

9. Minakata, M., Saito, S., and Shikata, M., 1979, J. Appl Phys., 50, 3063-3067.

10. Minakata, M., Saito, S., Shikata, M., and Miyazawa, S., 1975, J. Appl. Phys., 49, 4677-4682.

11. Esdaile, R.J., McLachlan, A.D., and De La Rue, R.M., 1980, Topical Meeting on Integrated and Guided Wave Optics, Incline Village, Nevada, WB-3.

12. Vassel, M.O., 1971, J. Opt. Soc. Am., 61, 166-173.

13. Esdaile, R.J., 1978, Appl. Phys. Lett., 33, 733-734.

14. Smith, D.S., Riccus, H.D., and Edwin, R.P., 1976, Opt. Comm., 17, 733-734.

15. Crank, J., 1975, The Mathematics of Diffusion, Oxford University Press, New York.

16. Hocker, G.B., and Burns, W.K., 1977, Appl. Opt., 16, 113-118.

17. Taylor, H., 1976, IEEE J. Quantum Electron., QE-12, 748-752.

18. Chen, F.S. 1969, J. Appl. Phys., 40, 3389-3396.

19. Glass, A.M., 1978, Opt. Eng., 17, 470-479.

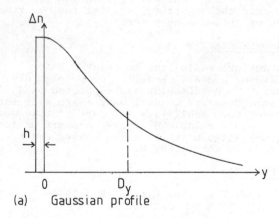

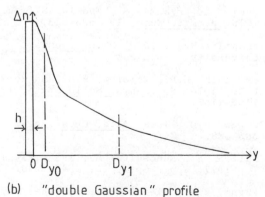

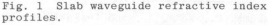

Fig. 1 Slab waveguide refractive index profiles.

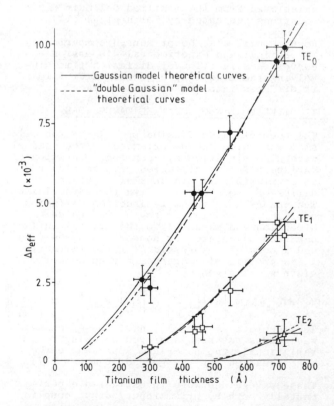

Fig. 2 Theoretical and experimental dispersion curves for slab waveguides diffused for 18 hours at 1000°C.

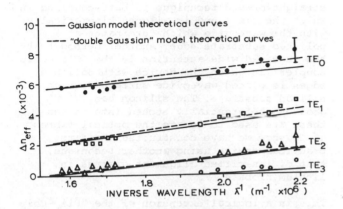

Fig. 3 Theoretical and experimental dispersion curves for slab waveguides as a function of optical wavelength. (Titanium thickness = 520 Å, 18 hours diffusion at 1010°C.)

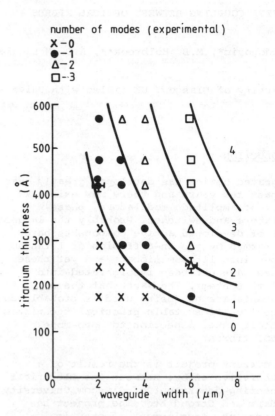

Fig. 4 Experimental modes and theoretical cut-off curves for channel waveguides. (8 hours diffusion at 1000°C)

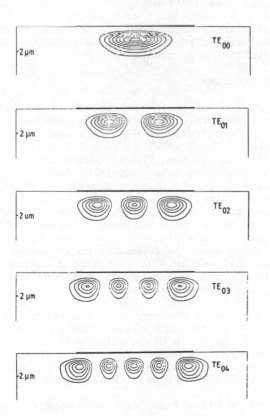

Fig. 5 TE_{00}-TE_{04} modes of an 8 μm wide channel waveguide. (575 Å Titanium, diffused for 8 hours at 1000°C.)

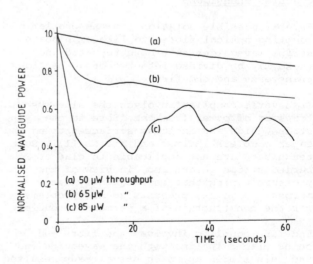

Fig. 6 Effect of photorefractive damage as a function of initial throughput power (10 μm wide channel waveguide)

END-FIRE COUPLING BETWEEN OPTICAL FIBRES AND STRIPE WAVEGUIDES

I. Andonovic*, M.B. Holbrook**, A.D. McLachlan***

University of Glasgow, UK (*Also with University of Strathclyde, **Also with Barr & Stroud Ltd, UK)

INTRODUCTION

Integrated optics has developed greatly over the past few years and there now exists a number of amplitude modulators, phase modulators and switches. Probably the largest range of developed active components have been formed by the in-diffusion of titanium stripes into lithium niobate and yet these devices have not been incorporated into complete systems. The fact that these components are not being used is probably due to the lack of suitable practical techniques for their interconnection to mono-mode optical fibres.

This current project is the result of a collaborative effort between the Electrical Engineering department of Glasgow University and Barr and Stroud Ltd. The project has two aims: the development of practical coupling techniques so that stripe waveguide devices may be incorporated into prototype systems where their performance as components may be evaluated and the development of longer-term solutions which are amenable to automated production.

The techniques for coupling between titanium in-diffused stripe guides will now be reviewed and three schemes which are receiving practical attention at Glasgow will then be introduced. This will be followed by details of their current state of development at Glasgow and the projected future work.

COUPLING TECHNIQUES

Several possible solutions to the problem of coupling optical fibres to lithium niobate stripe waveguides have been reported and these may be divided into two broad sections - transverse and end-fire.

Transverse coupling involves the sideways transfer of power from the fibre to the stripe guide and includes surface-grating and taper couplers (Bulmer and Wilson (1)). Such techniques are not applicable to clad or buried stripe guides and, in view of the pressure tuning that is frequently used across the joint, presents severe problems for the construction of a permanent coupler.

End-fire coupling involves the fibre end being placed in line with the waveguide end and this simple approach has already resulted in "potted" prototypes (Noda et al (2)). It is end-fire techniques that are being considered here.

***Now with Racal-MESL Ltd, UK

There are three possible approaches to the end-fire problem. Conceptually the most straightforward technique is butt-coupling in which the fibre end is aligned kinematically with the waveguide which terminates at a polished substrate edge. Another method that has received wide attention is the "Flip-Chip Coupler" in which the device with polished edges is placed waveguide surface down on a silicon substrate. The silicon has 'V' grooves preferentially etched into it and these are used to locate the optical fibres. At Glasgow we have constructed the first practical device using another technique, that of ion-milled alignment grooves in lithium niobate.

This is a logical extension of the "Flip-Chip Coupler" and consists of using ion milling to form grooves at the end of the waveguides in the lithium niobate itself. No edge polishing is needed here as the waveguide is terminated by an ion-milled wall. All the necessary alignment is carried out at the photo-lithographic stage under a mask aligner and the technique appears to be amenable to automated production.

All the above techniques are under practical investigation at Glasgow as is indicated in the following sections.

Butt-Coupling

Figure 1 shows an initial prototype utilising this technique. The stripe waveguide was formed by diffusing a 8 μm wide stripe of titanium, 220 Å thick at 1000°C for 9 hours in lithium niobate. The diffusion took place in a closed crucible containing lithium niobate powder to prevent out-diffusion of lithium oxide and the resulting guide supported a single transverse mode. The edges of this device were polished by using shellac to bond it to a piece of quartz and then by polishing down the composite on a diamond paste followed by Syton. Polishing continued until the edge exhibited no scratches or chips at x400 magnification.

As is shown in the schematic the waveguide was then placed on a glass block support in a perspex box which had a sliding side. 2 mm holes were drilled in the box, one in the removable side and one opposite, both being approximately in line with the waveguide.

The fibre had a 7 μm core and supported three modes at 0.6328 μm which was the experimental wavelength. The end was prepared by cleaving and then cemented into a 2 cm length of hypodermic steel capillary, of 1 mm O.D., with the end of the fibre projecting $\sim$ 10 μm beyond the end of the tube. The hypodermic was in turn supported in a glass capillary mounted on a XYZθ kinematic adjustment. The jointing operation consisted of removing the sliding side from the box and focussing a x40 objective on the output edge of the waveguide. The sliding side was required in order to bring the objective up to this output edge. Light was then launched into the fibre from a 2 mW HeNe laser and the fibre end encased in its steel capillary offered up to the polished input edge of the waveguide through the hole in the side of the box. Adjustment of the fibre in its kinematic mount and levelling of the device which was mounted on a prism table enabled light to be coupled into the waveguide. The operation was monitored by observing the image of the output edge on a screen and by a binocular microscope mounted above the coupling jig.

Once the alignment had been optimised the fibre was brought into contact with the waveguide edge and a drop of U.V. curing cement applied so as to form a fillet from the metal capillary, across the fibre end and onto the waveguide substrate. The use of U.V. curing adhesive admitted the possibility of further optimisation if drift occurred during the addition of cement, although this was usually unnecessary. After the fibre had been cemented to the waveguide the over-sized drilling in the box wall was filled with a viscous epoxy in order to rigidly support the fibre assembly.

The efficiency of the fibre joint was 6 dB, which is uncorrected for Fresnel reflection and waveguide losses. After the first joint had been formed the objective was removed and the side wall replaced and cemented into position. An output fibre was then located and fixed in the same manner as the input fibre. After all joints had set the box was filled with a slow-curing casting epoxy resin.

We expect to improve on the figure of 6 dB by using mono-mode fibre at 0.6328 μm and waveguide dimensions designed to maximise the field overlap between the fibre and waveguide modes. We also expect to increase the efficiency by using anti-reflection coatings.

Flip-Chip Coupling

This approach has been investigated by several workers (Hsu and Milton,(3)), Bulmer and Sheen (4)). As has been stated above the original intention was to use registration marks for alignment. It was found by some workers however, (4) that this procedure did not produce efficient coupling and they had to revert to introducing an adjustment, by means of a tapered alignment fibre located in a 'V' groove at right angles to those accommodating the launching fibres. The incorporation of such an adjustment immediately makes the arrangement prone to vibration.

In our work we have used silicon masked with 3000 $\overset{\circ}{A}$ of thermally grown silica. The silica is patterned by exposing, developing and baking a 0.4 μm layer of 1350J resist and subsequently etching the silica through windows in the resist with buffered hydrofluoric acid. Using this silica mask the silicon was then preferentially etched by refluxing in a I.P.A./NaOH/H_2O etch. Providing that the 'V' groove is aligned to $\sim 1^{\circ}$ with the $\angle 100 >$ direction excellent results were obtained which indicate that sufficiently accurate grooves can be etched with contemporary standards of photolithography. Preliminary work also indicates that the standards of uniformity and concentricity of commercially available fibre is within specification and not a limiting factor. Thus we expect the "Flip-Chip" arrangement as originally conceived (3) to play an important part in prototype work.

Ion Beam Milled Alignment Grooves in Lithium Niobate

Lithium niobate is not amenable to chemical etching and this technique cannot be used to etch fibre alignment grooves into the waveguiding substrate. It has been established however that ion-milling of the material is possible with high resolution both in the width and depth of the etched pattern (Kaminow and Ramaswamy (5)). This process cannot pattern a substrate to any great depth however and thus any fibre that is to be located in an ion-milled groove would have to be etched down close to the core. The use of etched down fibres would also have the advantage of permitting coupling to closely spaced waveguides, such as at the output of 'Y' junctions. In the experiments described below a sandwich ribbon fibre has been used which consists of a planar substrate supporting an exposed core (Laybourn and Millar (6)). Shallow grooves can be used to locate the cores of these fibres without reverting to etching the fibre.

Whichever fibre is used the groove depth will have to be in excess of $\sim$ 3 μm and this requirement brings with it masking problems. Carbon has an extremely low sputter yield, 12% that of lithium niobate, and would seem to be an ideal choice for a masking material. Initially we used a carbon-arc-deposited film delineated by lift-off of photo-resist stripes. The resulting mask was poor (Figure 2, Top) but it did permit the construction of a demonstration device (Figure 3).

In this device 8 μm stripe guides were first formed by diffusing in 200 $\overset{\circ}{A}$ thick stripes of titanium at 1000°C for 9 hours. Then using a positive mask square-ended AZ1350J resist ridges 8 μm wide, 0.4 μm thick and 2 mm long were deposited in line with the waveguides. The sample was then coated with $\sim$ 1 μm of carbon and windows opened in this film by lift-off. Ion etching was then carried out for 2 hours at a current density of 0.8mA/cm^2 which formed 4 μm deep grooves. Subsequently the excess carbon was removed by plasma ashing in oxygen. A short length of sandwich ribbon fibre with a 8 μm x 3 μm core was then placed in this groove by kinematic adjustment and bonded into position with U.V. curing cement. Light at 0.6328 μm was then coupled into the guide by a rutile prism and left via the sandwich ribbon fibre core.

This device illustrates the practicality of the idea, however the waveguide was very lossy due to etch-pits and thus our recent development work has centred around the production of good carbon masks.

We have developed a technique for the high temperature deposition of carbon onto a suitable semiconductor buffer layer (1000 Å thick) which allows the production of a relatively thick, hard and adherent film. These layers have good uniformity and small grain size. These films were patterned by reactive plasma etching in oxygen through aluminium masks. It is hoped that the quality of mask shown in the lower part of Figure 2 will allow us to construct more efficient couplers.

CONCLUSION

It appears at present that one of the principal applications for stripe waveguide devices will be as optical processors for fibre optic sensor systems. Designs for such components already exist (Papuchon and Puech (7)) however they cannot be utilised until the problem of linking the processor permanently to the sensing element, which consists of mono-mode fibre, has been solved. It is hoped that the work reported above will solve this problem.

The authors wish to thank Professor J. Lamb and the members of his group particularly Dr. R. De La Rue and Dr. S. Wright, Mr. R. Hutchins and Mr. G. Boyle. Also the authors wish to extend their gratitude to the members of the physics department at Barr and Stroud Ltd. particularly Mr. B. Monachan, Mr. C. Scott and Dr. R. Corbett, and the crystal production department for the supply of lithium niobate.

A. McLachlan was supported by an S.R.C. C.A.S.E. award with the Plessey Co. Ltd. and facilities made available at the Plessey (Caswell) Ltd. Allan Clarke Research Centre are gratefully acknowledged.

References

1. C.H. Bulmer, M.G.F. Wilson; I.E.E.E. J. Quant. Elec. Q.E.-14, 741-749, '78.

2. J. Noda, O. Mikami, M. Minakata, M. Fukama; App. Opt., 17, 2092-2096, '78.

3. H.P. Hsu, A.F. Milton; Elec. Lett., 12, 404-405, '76.

4. C.H. Bulmer, S.K. Sheem, R.P. Moeller, W.K. Burns; App. Phy. Lett., 37, 351-353, '80.

5. I.P. Kaminow, V. Ramaswamy; App. Phy. Lett 24, 622-624, '74.

6. P.J.R. Laybourn, C.A. Millar; Elec. Lett. 10, 175-176, '74.

7. M. Papuchon, C. Puech; SPIE, 157, 218-223, '78.

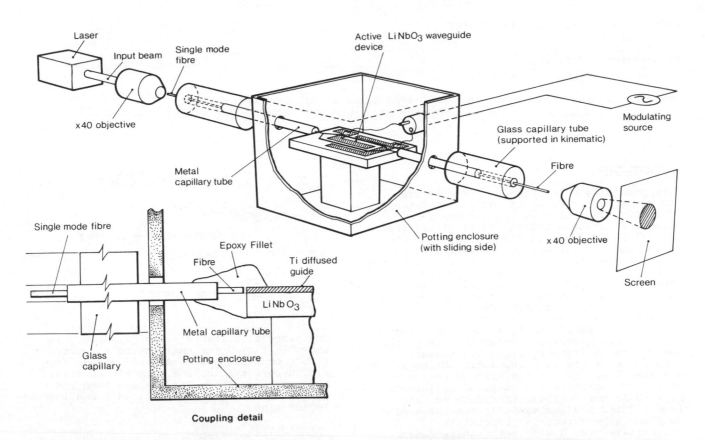

Fig.1 Schematic diagram of butt coupling arrangement.

Fig.2 Top. Carbon mask produced by arc deposition and lift off (1 μm markers). Bottom. Oxygen etched film deposited on keying layer. (1 μm markers).

Fig.3 Top. Ion etched groove in LiNbO₃ (1μm. markers). Bottom. Sandwich ribbon fibre placed in above groove and epoxied into position. (10 um markers).

TEMPERATURE COMPENSATED PRISM COUPLER

F. Auracher and R. Keil

Research Laboratories of Siemens AG, D-8000 Muenchen 83, West Germany

INTRODUCTION

Very promising waveguide devices make use of the electrooptic or acoustooptic properties of $LiNbO_3$. Typical examples are Bragg-deflectors using out-diffused or Ti-indiffused optical waveguides. Coupling light into these waveguides is most easily achieved by means of prism couplers as shown in Fig. 1. Recently high input coupling efficiency has been obtained using prism couplers with a tapered gap (1,2); also out-diffused waveguides with high threshold for optical damage have been fabricated (1) thus paving the way to practical waveguide devices. One remaining obstacle for practical implementation of these devices is the rather strong temperature dependence of the coupling angle. Using a parallel laser beam a temperature change of only 5 °C decreases the coupling efficiency by 3 dB in our devices. Therefore we have investigated and successfully demonstrated a coupling scheme that is temperature insensitive over a wide temperature range.

CONDITIONS FOR TEMPERATURE COMPENSATION

We require that the coupling angle α defined in Fig. 1a and given by

$$\alpha = \arc \sin\left[n_p \cdot \sin\left(\gamma -\arc \sin \frac{n_{eff}}{n_p}\right)\right] \quad (1)$$

is temperature independent. Assuming a linear variation of all relevant refractive indices with temperature we obtain for the suitable temperature coefficient a_p of the refractive index n_p of the prism

$$a_p = a_s\left(\frac{n_{eff_o}}{n_{p_o}} + \frac{1}{\tan\gamma}\cdot \sqrt{1-\left(\frac{n_{eff_o}}{n_{p_o}}\right)^2}\right), \quad (2)$$

where a_s is the corresponding temperature coefficient of the $LiNbO_3$ substrate, n_{eff} the effective refractive index of the waveguide and the subscript "0" refers to room temperature.

The shaded area in Fig. 2 shows the range of suitable temperature coefficients of the coupling prism for an out-diffused waveguide with $n_{eff_o} = 2,201$ ($\lambda = 0,633 \ \mu m$). It can be seen that none of the better known materials is suited. However, adding a second prism (Fig. 1b) with a vastly different temperature coefficient can solve the problem as will be shown. The required refractive index n_{c_o} and its temperature coefficient a_c are given by

$$n_{c_o} =\left[\left(\sqrt{n^2_{p_o} -n^2_{eff_o}}\cdot \frac{\sin\gamma}{\sin(\gamma -\beta)} + \frac{\sin\alpha}{\tan(\gamma -\beta)} - \right. \right.$$
$$\left. \left. - n_{eff_o} \frac{\cos\gamma}{\sin(\gamma -\beta)}\right)^2+\sin^2\alpha\right]^{1/2}$$

$$a_c = \frac{\sqrt{n^2_{p_o} -n^2_{eff_o}}\cdot \sin\gamma+\sin\alpha\cdot\cos(\gamma -\beta) -n_{eff_o}\cdot \cos\gamma}{n_{c_o}\sin^2 (\gamma -\beta)}\cdot$$
$$\cdot\left(\frac{a_p n_{p_o} -a_s n_{eff_o}}{\sqrt{n^2_{p_o} - n^2_{eff_o}}}\cdot \sin\gamma -a_s\cos\gamma\right) \quad (3)$$

The range of suitable prism parameters is shown in Fig. 3 for the case of a c-cut $LiNbO_3$ coupling prism. As can be seen from Fig. 3 polystyrole is a suitable material, the more common organic glasses PMMA and polycarbonate should also yield a considerable improvement.

EXPERIMENTAL RESULTS

A polycarbonate prism coupler was fabricated and attached to the $LiNbO_3$ coupling prism by means of an immersion liquid. This material has good mechanical and optical properties and is readily available. The $LiNbO_3$ substrate with the attached prisms was mounted in a small, closed specimen chamber that could be heated or cooled thermoelectrically. He-Ne laser light could be coupled into and out of the chamber through optical windows. The chamber was mounted on a goniometer table enabling us to measure the input coupling angle accurately. The measured input coupling angle as a function of temperature is shown in Fig. 4. The line marked (a) shows the results for the $LiNbO_3$ prism alone; the one marked (b) clearly demonstrates the significant improvement due to the additional polycarbonate prism. For polystyrole the expected performance lies along line (c). Experimental results with a polystyrole prism will also be reported.

CONCLUSION

We have for the first time demonstrated a temperature insensitive prism coupler for diffused waveguides in $LiNbO_3$. This improvement brings waveguide devices based on $LiNbO_3$ one step closer to practical implementation.

REFERENCES

1. Sarid, D., Cressman, P.J. and Holman, R.L., 1978, Appl. Phys. Lett., 33, 514-515

2. Auracher, F., Keil, R. and Zeitler, K.-H., 1981, Siemens Forsch.- u. Entw.-Ber., 10, 44-47

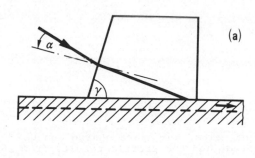

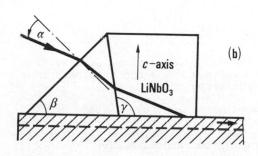

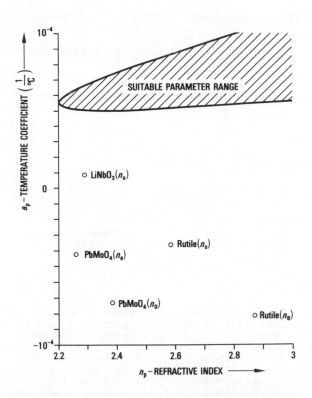

Figure 1 Prism coupling of laser beam into planar waveguide, a) with single coupling prisms, b) with combination of two prisms

Figure 2 Parameter range of coupling prism for temperature intensitive input coupling in coupling scheme according to Fig. 1a

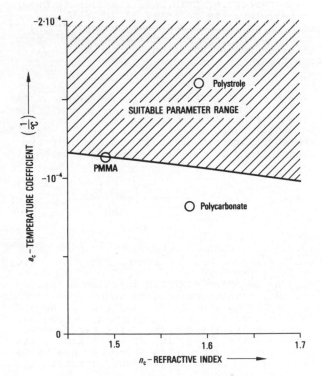

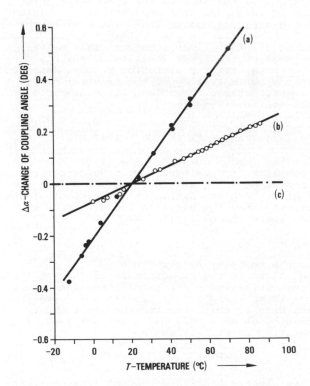

Figure 3 Parameter range for second prism for coupling scheme according to Fig. 1b

Figure 4 Variation of input coupling angle with ambient temperature, (a) for $LiNbO_3$ prism alone with $\gamma = 75°$, (b) for a combination of a $LiNbO_3$ ($\gamma = 105°$) prism and a polycarbonate prism ($\beta = 40°$), (c) expected behaviour for polystyrole prism

FIELD-INDUCED INDEX PROFILES OF ION EXCHANGED MULTIMODE STRIP WAVEGUIDES

D. Ritter, H.-J. Lilienhof, E. Voges

FernUniversität Hagen, Germany

INTRODUCTION

In multimode fiber-optic systems optical branching networks are of interest for data distribution. The fabrication of planar branching components by photolithographic techniques has the well known advantages of reproducibility and batch fabrication. Multimode components have been formed preferably by ion exchanged processes in glasses /1-3/. One of the unsolved problems is the matching of the profils and the geometries of the strip waveguides to connecting step-index or graded index fibers in order to achieve a high coupling efficiency. This requires the determination of ion exchanged index distributions in dependence on the technological parameters. Here, we show that the profiles of ion exchanged strip waveguides are mainly influenced by the applied external field distribution, and that a precise calculation is possible. Theoretical and experimental results are presented.

THEORY

The index profiles of silver ion exchanged waveguides obtain from the concentration of the Ag-ions /4/. This concentration is calculated assuming sodium ions (index N) as the only mobile ion species within the glass, and using diluted silver (index A) salt melts. Because of the lower mobility of the Ag-ions the field assisted diffusion process depends on the concentration of the silver ions due to space charge formation. It is governed by the nonlinear equation

$$\frac{\partial c}{\partial t} + \frac{1}{1-(1-M)c} \ \mu_A \vec{E} \ \mathrm{grad} c = \frac{D_A}{1-(1-M)c} \ \Delta c \qquad (1)$$

for the concentration of the silver ions normalized to the equilibrium Na-concentration c_N of the glass. Here, μ_A and D_A are the mobility and the diffusion coefficient of the silver ions, the electric field $\vec{E}$ includes the space charge fields, and

$$M = \mu_A/\mu_N < 1 \qquad (2)$$

is the mobility ratio. Since eq. (1) cannot be solved in the two-dimensional case we first investigate approximate one-dimensional solutions where c is a function of depth y and time t, only. For the special boundary conditions at $y = 0$: $c = c_o$ and $\partial c/\partial y = 0$, a solution of eq. (1) is /5/

$$c/c_o = \{ 1 + \exp [v \ c_o \frac{1-M}{D_A} \ (y-vt)] \}^{-1} \qquad (3)$$

with a concentration dependent front velocity

$$v = v_o \frac{1}{1-(1-M)c_o} \qquad (4)$$

where $v_o = i_o/c_N$ depends on the current density i_o. The solution (3) only applies for large profile depths because of the boundary conditions. For low silver concentrations

(diluted melts) the space charge can be neglected, and the profile results from /6/

$$c/c_o = \frac{1}{2} \left\{ \mathrm{erfc} \ \frac{y-v_A t}{2 \ \sqrt{D_A t}} + \exp \ \frac{v_A y}{D_A} \right.$$

$$\left. \cdot \ \mathrm{erfc} \ \frac{y+v_A t}{2 \ \sqrt{D_A t}} \right\} \qquad (5)$$

where the front velocity v_A obtains from the applied voltage U and the thickness d of the glass substrate

$$v_A = v(c_o \to 0) = \mu_A U/d \qquad (6)$$

In the most simple solution the diffusion is neglected, too, and we arrive at

$$c/c_o = \sigma (v_A t - y) \qquad (7)$$

where σ is the unit step function. The solutions (3), (5), (7) are compared in fig. 1 for the parameters $D_A = 2.6 \cdot 10^{-15} \ m^2/s$, $t = 15$ min, $U/d = 300 V/mm$. The simple step solution well applies for large depths of the profile. Therefore, this approximation is used to calculate the profiles of strip waveguides for configurations as shown in fig. 2. The shape of the uniform index distribution then is given by the position vector of the silver ion front

$$\vec{r}_F = a \ \vec{e}_x + \int_0^{t_1} \mu_A \vec{E} \ dt, \ -x_w/2 \leq a \leq x_w/2 \qquad (8)$$

for an exchange time t_1. The electric field for the configuration of fig. 2 is equal to the static field of the triplate transmission line

$$E = E_x + jE_y = j \ \frac{\pi}{2} \ \frac{U}{dK'(k)} \cdot \left[\frac{\tanh^2 u - 1}{\tanh^2 u - k^2} \right]^{1/2} \qquad (9)$$

with

$$u = \pi (x - j y)/2d \ , \ k = \tanh \pi x_w/4d \qquad (10)$$

and the Jacobian elliptic function $K'(k)$. Fig. 3a, b shows calculated index profiles for two widths x_w and several applied voltages U (parameters: $T = 616K$, $t_1 = 30$ min, $d = 300 \mu m$). These results are compared with experimental index profiles (fig. 3c, d) which are fabricated for the above parameters and are visualized by Nomarsky interference contrast.

EXPERIMENTAL RESULTS

The optical strip waveguides are fabricated in borosilicate glasses by a field-assisted migration of silver ions from 10mol% AgNO$_3$/ 90mol% NaNO$_3$ melts. The cathode is formed by an evaporated gold layer (fig. 2), the mask consists of sputtered titanium, SiO$_2$ or Si$_3$N$_4$

with equal results. The two-dimensional index
distributions are precisely evaluated with
the aid of an optical multichannel analyzer
/7/. Fig. 4a shows a two-dimensional plot of
the measured profile, and fig. 4b the corresponding Nomarsky contrast. Beveled samples
are used for an enlarged resolution in depth.

A comparison between theoretical and experimental profiles shows, in general, that the
profiles can accurately be calculated from
the technological parameters. Most important
are the depth of the profile and the broadening due to side diffusion. Fig. 5a shows
that the depth of the profile increase for
smaller widths x_w of the window due to the
field concentration as expressed in eq. (9),
whereas the side diffusion is almost constant
as depicted in fig. 5b. Because of this
effect, the depth and broadening of the profile are nonlinear functions of the exchange
time t_1 (fig. 5c, d).

In addition, the index profiles obtained by
a second ion exchange in pure $NaNO_3$ melts
may be calculated and fitted to the profiles
of connecting fibers. These results will be
presented together with branching components
fabricated by field-assisted ion exchanges.

ACKNOWLEDGEMENTS

This work has been supported by the Deutsche
Forschungsgemeinschaft (DFG).

REFERENCES

/1/ G.L. Tangonan et.al., Opt. Comm. Con-
     ference Amsterdam 1979, paper 21.5

/2/ O.G. Ramer, Appl. Optics, 19 (1980),
     1294

/3/ J. Viljanen, M. Leppihalme, Appl. Phys.
     24 (1981), 61

/4/ M.L. Huggins, Kuan-Han Sun, Jour.Am.Cer.
     Soc. 26 (1943), 4

/5/ M. Abou-el-Leil, A.R. Cooper, Jour.Am.
     Cer.Soc. 62(1979), 390

/6/ T. Kaneko, H. Yamamoto, 10th Intern.
     Congress on Glass, Part I, (1974), 8 - 79

/7/ H.J. Lilienhof et.al, Opt. Comm. 75
     (1980), 49

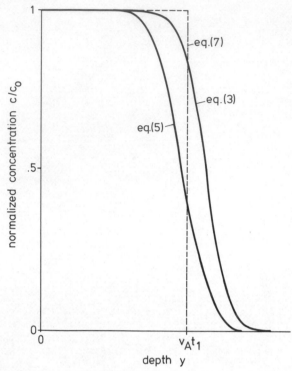

Figure 1 Approximate solutions of one-dimensional Ag-ion concentration profiles. The solution (3) applies for $c_0 = 0.3$

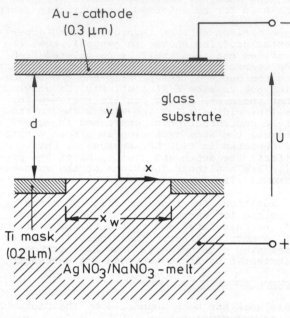

Figure 2 Configuration of a field-assisted ion exchange for optical strip waveguides

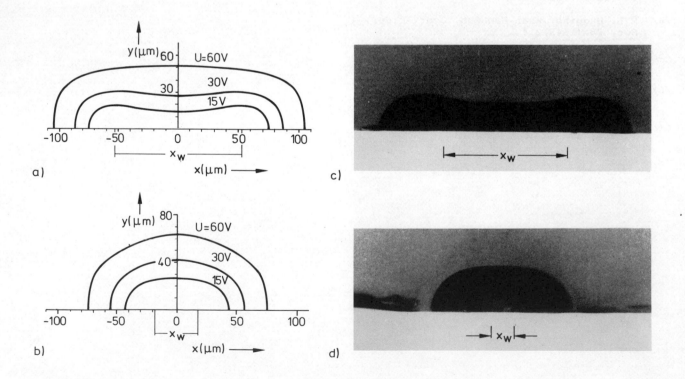

Figure 3 Calculated index profiles (a), (b) and experimental profiles visualized by Nomarsky interference contrast (c), (d).

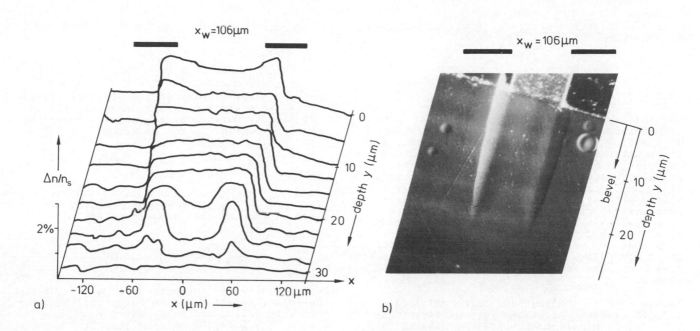

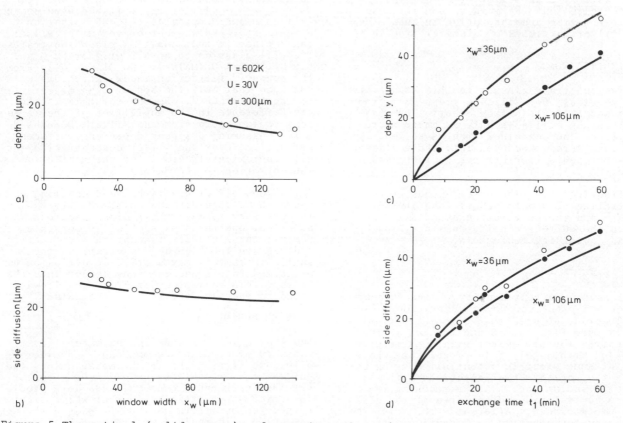

Figure 4 Index profiles scanned by an optical multichannel analyzer on a beveled surface (a) and related Nomarsky contrast (b) for an ion exchange with T = 615K, t_1 = 30 min, U = 30 V, d = 300 μm

Figure 5 Theoretical (solid curves) and experimental (● ○) results for the profile depth and the side diffusion in dependence on the width x_w of the window (a), (b) and exchange time t_1 (c), (d)

ANALYSIS OF LOSS IN ION EXCHANGED GLASS WAVEGUIDES

J. Viljanen and M. Leppihalme

Technical Research Centre of Finland, Finland

INTRODUCTION

The ion migration technique in various forms offers several advantages in the fabrication of coupling elements for multimode fibres. In the coupler structure the number of inputs and outputs can be varied within the photolithographic process of the diffusion mask fabrication, and because the guide preparation is a planar process it is also possible to fabricate several elements on the same substrate glass, which is then cut to separate the couplers. Since diffusion smoothens irregularities that are smaller than the guide diameter these waveguides are potentially of the lowest loss among the strip waveguides in integrated optics.

Though several studies have been published on the ion exchange technique very little attention has been paid to the loss mechanisms in these waveguides. In this work we have analysed the losses in silver ion exchanged waveguides. Although we have used here the two step migration technique with thin film Ag-layer as the ion source the origin and formation of loss would be the same if a molten salt ion source with or without electric field were used. Three different types of substrate glasses were systematically treated. The results indicate that the main reason for loss is the scattering and absorption in submicroscopic silver particles of diameter from 10 to 20 nm.

GUIDE FABRICATION

The three substrate glasses in this study were soda-lime glass, borosilicate glass 0211 from Corning and BK-3 glass from Schott. The soda-lime glass was available only as cheap microscope slides and contained iron as an impurity. Silver ions were introduced into these glasses from evaporated silver stripes on the glass surface (Chartier (1)).

The refractive index increase depends on the concentration of mobile sodium ions in the substrate and the penetration speed dx/dt depends on the current density passing through the glass according to formula

$$\frac{dx}{dt} = \frac{i}{F \cdot c} , \qquad (1)$$

where i is the current density, F the Faraday constant and c the sodium concentration. As back electrode we used thin evaporated aluminium film, see Fig. 1a. Since the ion source was a narrow (30 μm) silver stripe pattern it was coated with a gold film to hold the whole stripe at same electric potential. During the process the negative back electrode attracts the mobile sodium ions. After reaching it they are neutralized and gathered at the glass aluminium interface. Afterwards this could be verified by touching the back surface with a wet pH-indicator paper. Because the mobility of silver is smaller than the mobility of sodium the resistivity of the glass increases during the migration. This decreases the current density during the migration. Using the notations of Fig. 1a we can write

$$i = \sigma_{Na} \frac{V - V_x}{d - x} = \sigma_{Ag} \cdot \frac{V_x}{x} , \qquad (2)$$

where σ_{Ag} and σ_{Na} are the conductivities in the exchanged and unexchanged parts of the glass. From eqs. (1) and (2) we get

$$\frac{di}{dt} = \frac{i^2}{F c d} \left(\frac{\sigma_{Na}}{\sigma_{Ag}} - 1 \right) . \qquad (3)$$

From the observed current curve in Fig. 1b for the soda-lime glass we can calculate the conductivity ratio. With i = 131 μA/cm^2, d = 0.1 cm, c = 5.1×10^{-3} mole/cm^3, di/dt = -2.9×10^{-3} μA/scm^2 we get for the soda-lime glass σ_{Na}/σ_{Ag} = 9.1. This value is in agreement with the reported mobility ratios of the two ions (Doremus (2)). The maximum refractive index increase at the surface was measured using the prism coupler method and assuming a step like index distribution (Ulrich and Torge (3)). For soda-lime glass the obtained Δn was 0.081, for BK-3 0.03 and for 0211 glass 0.042. Earlier it had been shown (Viljanen and Leppihalme (4), (5)) that an almost total replacement of sodium by silver takes place in the soda-lime glass substrates. The 0211 glass contains 6.2 wt.% of Na$_2$O and 6.9 wt.% of K$_2$O (Maissel and Glang (6)). From the measured refractive index increments we conclude that only the sodium ions migrate during the process. Reason for this is the big difference in the mobilities of the three ions sodium, potassium and silver (Doremus (7)). Assuming a linear dependency between refractive index and silver concentration one gets approximately 4.5 % for the mobile sodium concentration in the BK-3 glass.

For a good efficiency in but-end coupling between the waveguides and multimode fibres it is necessary to make their cross sections and numerical opertures as similar as possible. For all three glasses the obtained refractive indices are too high and penetrations too shallow. That is why a drive-in diffusion was performed to diffuse the ions deeper into the glass. This took place at 370°C and it was found to increase the loss of the guides.

LOSS MEASUREMENTS

The studied guides were approximately 50 μm wide, 20 μm deep and 40 mm long. In order to eliminate stray light in the measurements the guides were branchless and had a 90° bend after the input end. When the ends were studied in a microscope it was seen that white light turned yellow or orange when it passed through the guide. To analyse the loss mechanisms we measured the spectral dependency of attenuation using a prism monochromator and silicon

and germanium photodetectors. Before measurement the edges of the substrate were painted
black and the glass was placed between two
dark acrylate plates with glycerol as an immersion liquid between the surfaces. For the
excitation we used a 10x achromatic microscope objective. The photodiode in turn was
mounted directly against the end of the waveguide. The absolute value of the loss was
measured with a HeNe-laser (633 nm) and
scanning along the guide with a small area pin
diode. In Figs. 2 and 3 the spectral losses
are given in dB/cm. The loss around 400-500 nm
wavelength is caused by the scattering and
absorption in small submicroscopic silver
particles precipitated inside the guide. The
scattering and absorption spectra of small
metal spheres have been theoretically analysed
by Mie (8), and for silver experimentally studied by several authors (Doremus (9), Yamamoto
(10), Bamford (11)). The position of the loss
maximum is determined by the optical coefficients of metallic silver and the shape and
size of the particles.

Before precipitation to silver particles is
possible, a fraction of the ions must be neutralized to atomic silver. To find out the
effect of the surface electrode three different migration voltages were used. No remark-
able changes were observed in the loss spectra when the voltage was increased from 5 to
35 volts. However the loss was smaller when
the first migration step was performed without the gold electrode on the silver stripe.
At the guide edges the neutralization hap-
pens through the blocking anode current of
the gold electrode. This current flows
through the space charge layer formed by the
depletion of sodium ions from the vicinity
of the positive gold electrode. The electric
field in this depletion region is very high
(high ion concentration inside the glass)
and it can break the chemical bonds between
oxygen and silicon atoms in the glass network (Carlson (12)). In this way also oxygen
can migrate and for each removed oxygen two
neutral silver atoms are left into the waveguide. The formation of silver colloids can
be promoted also by exposure to ultraviolet
light and addition of polyvalent oxides, like
iron or cerium oxides (9), (10). This has
been used in photochromic and yellow stained
silver glasses. Another method that has been
observed to enhance the silver particle formation is the application of external pressure (10). During guide fabrication considerable stress arises into the glass when sodium
ions are replaced by bigger silver ions. The
two possible explanations for the additional
loss maximum at 600 nm in Fig. 2 curve B are
the presence of Ti^{3+} ions (Bates (13)) or
elongated silver particles inside the guide
(Stookey and Araujo (14)).

During drive-in diffusion the particles grow
bigger, their absorption and scattering move
to longer wavelengths. At the same time the
colour of transmitted light changes from yellow to orange. However, it is difficult to
estimate the increase in colloidal silver
from the loss curves in Figs. 2 and 3. This
might be possible if the particles were
spherical (11). In both fabrication steps
the precipitation takes place at relatively
low temperatures, where the viscosity of
glass is high. This can lead to deformation
in the particle shape. As a matter of fact
Weyl (15) attributes the red colour in silver
glasses to the irregular shape of the particles. The dotted curve in Fig. 3 is the Fe^{2+}
absorption of the soda-lime glass.

ADDITIONAL LOSS AT BRANCHING POINTS

Lateral diffusion of the ions during fabrication can cause additional loss into waveguide
intersections and branching points. To examine
this type of attenuation a branching element
with three branching points was fabricated
into BK-3 glass. The loss was estimated by
measuring the total transmitted power from
the magnified image of the output ends in
Fig. 4b. using a 1 mm² silicon photodiode.
When the 2 cm long element in Fig. 4a was
excited with a HeNe laser beam through a
20x objective the total transmission percentage was 31 %. Taking into account also the
Fresnell reflections at the input and output
surfaces the maximum value for the excess
loss is thus 4 dB. This means less than 2 dB
per branching point since part of the power
is lost by Mie scattering and absorption in
the waveguides, Fig. 4c. Exact evaluation of
the excess loss would require the use of
optical fibres at the input and output ends
of the element.

CONCLUSION

Our measurements revealed that the main reason
for loss in silver ion exchanged optical waveguides is scattering and absorption in submicroscopic silver particles. The resonance
wavelength for smaller particles precipitated
during the electric field induced migration
is about 410 nm. During further heat treat-
ments the particles grow and their characteristic absorption maximum moves to longer wavelengths. In the near infrared the optical
loss is determined by the tail of this loss
mechanism. To produce low loss waveguides it
is necessary to use iron free substrate
glasses.

REFERENCES

1. Chartier, G.H., Jaussaud, P., de Oliveira,
   A.D. and Parriaux, O., 1978, Electronics
   Letters 14, 132-134.

2. Doremus, R.H., 1964, J. Phys. Chem., 68,
   2212-2218.

3. Ulrich, R., and Torge, R., 1973, Appl. Opt.,
   12, 2901.

4. Viljanen, J., and Leppihalme, M., 1980,
   J. Appl. Phys., 51, 3563-3565.

5. Viljanen, J., and Leppihalme, M., 1981,
   Appl. Phys., 24, 61-63.

6. Maissel & Glang, Handbook of thin film
   technology, McGraw-Hill, Inc. 1970, USA

7. Doremus, R.H., 1969, Phys. Chem. Glasses,
   10, 28-33.

8. Mie, G., 1908, Ann. Phys., 4, 377-445.

9. Doremus, R.H., 1965, J. Chem. Phys., 42,
   414-417.

10. Yamamoto, T., Sakka, S., and Tashiro, M.,
    1969, J. Ceram. Assoc. Japan, 77, 378-385.

11. Bamford, C., Colour generation and control
    in glass, Elsevier Scientific Publishing
    Comp. 1977.

12. Carlson, D., Hang, K., and Stockdale, G.,
    1972, J. Am. Ceram. Soc., 55, 337-341.

20

13. Bates, T., Modern Aspects of the Vitreous
    State, ed. J.D. Mackenzie 1962, vol. 2,
    pp. 195-254, Butterworths, London.

14. Stookey, S., and Araujo, R., 1968, Appl.
    Opt., 7, 777-779.

15. Weyl, W., 1951, Coloured Glasses, Soc. of
    Glass Technol., Sheffield.

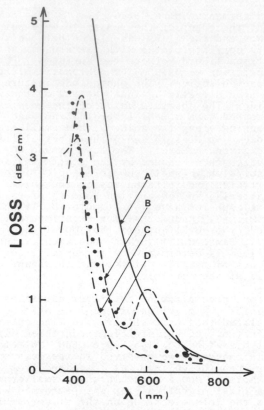

Figure 2 Loss before drive-in diffusion:
(A) 0211 with gold top electrode, (B) without gold, (C) etched 0211 glass, (D) BK-3

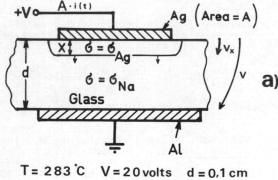

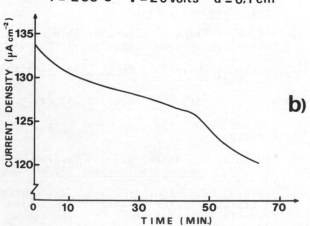

Figure 1 a) Setup for guide preparation and
b) current change during the process

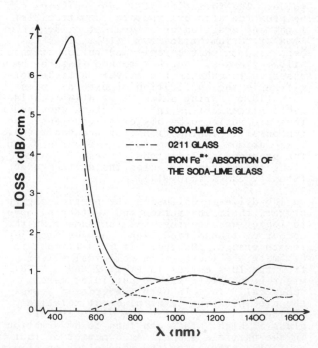

Figure 3 Loss after drive-in diffusion

21

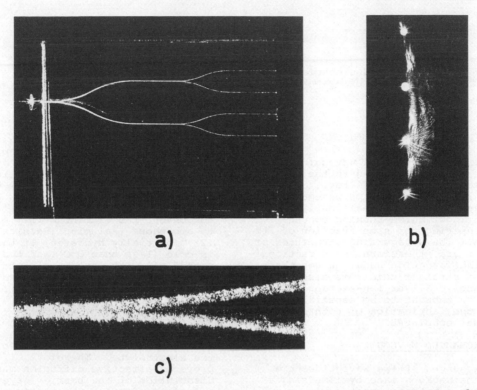

Figure 4 a) studied branching element, b) the four output ends and c) scattering from the
first branch

OPTICAL STRIPE WAVEGUIDES, DIRECTIONAL COUPLERS AND WAVEGUIDE BENDS FORMED BY SILVER ION EXCHANGE IN GLASS

R.G. Walker and C.D.W. Wilkinson

Department of Electronics and Electrical Engineering, University of Glasgow.

INTRODUCTION

This study of passive optical waveguiding structures is intended to lead to the design of a ring-resonator optical filter. Such a closed, circular loop of stripe waveguide is excited by means of directional coupling from an associated input guide running concentrically to the ring for some fraction of its circumference. The waveguiding structures of prime interest are, therefore the directional coupler and the waveguide bend, though the attenuation of straight guides is also of great importance. Silver ion-exchanged waveguides were thought to be especially suitable for the fabrication of both bends and directional couplers.

2. THE ION-EXCHANGED WAVEGUIDE

Silver ion-exchanged stripe waveguides are diffused channel guides made by immersing a soda-lime glass substrate, coated with a suitable masking material, into molten silver nitrate (melting point: 211°C). If apertures have been opened in the diffusion mask, sodium ions will leave the glass in these areas and silver ions will replace them. The resulting refractive index increase (~ 0.09 at the surface) is large enough to permit small bending radii.

Since the exchanging ions have the same charge but different mobilities (Ag^+ is less mobile than Na^+) an electric field is set up which equalises the ionic fluxes.

The two-dimensional non-linear diffusion equation which describes ion-exchange is solved numerically by digital computer using a 3-level finite difference algorithm. This yields a rectangular array of silver concentration values, representing the waveguide cross-section. Almost any 2-D geometry of ideal, insulating diffusion mask may be modelled.

Assuming that refractive index varies linearly with silver concentration, the Helmholtz scalar wave-equation may then be solved numerically using a variational technique. The wave amplitude function may be obtained, and also the propagation constants of all modes. The normalised propagation constant - the effective refractive index (n_e) - will be used throughout this paper.

3. EXPERIMENTAL STRIPE WAVEGUIDES

Two types of ion resistant masks have been used; a thin film of aluminium and an anodised aluminium film. Photolithographic techniques were used to open 2 to 20 μm apertures in a 0.1 μm film of aluminium deposited on microscope slides (n = 1.5125). The diffusion apertures were measured by observation of the Fraunhofer diffraction pattern of the illuminated aperture (Walker et al (1)).

After measurement of the waveguide widths, the aluminium masks were either used directly or completely converted to oxide by means of anodisation. This conversion is performed by making the aluminium thin film the anode in an electrolytic cell containing a 20% (V.V.) solution of sulphuric acid. It is possible to convert the aluminium film almost entirely to amorphous γ-alumina (Wernick and Pinner (2)) (partially hydrated) if the electrolyte is cold (less than 0°C) and the film partially masked off with photo-resist to provide a protected current path past the electrolyte surface. A constant voltage of 10-15 Volts was used.

Apart from a dielectrically compact barrier layer some 150 Å thick (2), the anodised film is porous due to its sparing solubility in the electrolyte. Therefore, if it is to prove an effective diffusion mask, the dissolution of the barrier layer must be halted immediately the anodic conversion is complete (rinsing briefly in water then in acetone is found to be effective).

Using either method of masking, ion exchange was carried out at 215°C. After exchange the mask was removed using phosphoric acid. The effective indices (n_e) of the guided modes were measured using a prism coupler and He-Ne laser light of wavelength (λ_o) 0.6328 μm. The accuracy of width and n_e measurements is estimated at ±0.04 μm and ±0.0002 respectively.

Propagation characteristics

The measured effective index values for stripe waveguides of different widths made on the same substrate using an aluminium mask are shown in Figure 1(a). No theoretical results are shown for comparison in this case as the modelling of the combined electrochemical and electrostatic boundary conditions for ion-exchange with a conducting mask is impractical. Not only does the conductivity of the aluminium affect the electrostatic situation, but an electrochemical potential is set up at the aluminium-silver nitrate interface (Chartier et al (3)).

The effective index values for different stripe widths using anodised masks are shown in Figure 1(b). The theoretical curves for a purely dielectric mask can be seen to agree well with experiment. An underirable feature of aluminium-mask waveguides is revealed when their surface is examined by a high power optical microscope (~ 1000 X) using reflected light. A cream-coloured stripe along each edge of the waveguide suggests the presence of metallic silver deposits. These are unaffected by nitric acid and are thus, presumably, buried. Anodised masks, by contrast, produce little or no such deposits and the resulting waveguides can have attenuation factors (dB/cm.) which are less than a quarter those of their metallic-mask counterparts - typically 3-4 dB/cm at 2 μm width for the anodised mask.

It is thought that ion-exchanging glass and aluminium together constitute an electrolytic cell when immersed in molten silver nitrate (3). If the metal develops a more negative potential, with respect to the melt, than does the glass surface (which behaves as a glass electrode, probably developing a net negative charge with respect to the melt) then ionic current will flow through the glass and silver will be deposited at the mask edges. This phenomenon is capable of explaining the metal-mask dispersion characteristic (Figure 1(a)) also.

4. DIRECTIONAL COUPLERS

Directional couplers have been made (using anodised masks) with guide widths between 2 and 2.6 µm, and guide separations between 1.85 and 1.3 µm. The spread is achieved by varying the photo-lithographic processing times, and the dimensions are obtained by measuring the double-slit interference pattern.

To measure the Coupling Coefficient (K) or Transfer Length (L), light (λ_O = 0.6328 µm) is coupled into one guide via a tapered section. A certain fraction transfers to the second guide before the guides diverge after running parallel for a distance s (the inter-action length). If an output prism is clamped onto the output arms (Figure 2), two separate m-lines result, whose intensities (I_1 and I_2) may be measured. According to standard directional coupler theory (Ramaswamy and Standley (4))

$$\frac{I_2}{I_1} = \frac{\sin^2(Ks)}{\cos^2(Ks)} = \tan^2(Ks) \qquad (1)$$

Thus: $Ks = m\pi + \tan^{-1}\left(\sqrt{\frac{I_2}{I_1}}\right)$; m = 0,1,2, (2)

and $\quad L = \pi/2K \qquad (3)$

The correct solution for Ks is obtained by comparing three different interaction lengths of 1 mm, 2 mm and 4 mm.

A theoretical transfer length may be obtained from the computer model using the measured dimensions. If the refractive index distribution of the parallel waveguide pair is generated (using the diffusion program) the wave-equation program will solve it as a single, composite guiding structure whose modes occur as nearly degenerate pairs - symmetric and antisymmetric. If these modes have effective refractive indices n_e^s and n_e^a respectively then,

$$L = \frac{\lambda_O}{2(n_e^s - n_e^a)} \qquad (4)$$

Table 1 below compares measured and computed transfer lengths for five couplers on one substrate. The guides were mono-mode (about 9 minutes diffusion).

TABLE 1 - Comparison between theoretical and measured transfer lengths of experimental directional couplers.

Coupler Dimensions Width (µm) ; separation (µm)		Transfer Length L (mm) Computed ; Measured	
2.34	1.53	5.25	5.44
2.27	1.53	4.65	4.61
2.36	1.51	5.1	4.13
2.3	1.47	4.05	5.3
2.44	1.33	3.15	3.43
±.05	±.05	±15%	-

5. EXPERIMENTAL CURVED WAVEGUIDES

180° ("hairpin") bends of radii 300, 200, 150, 125, 100, 85 and 70 microns have been made in stripe waveguide of 2 µm width. They may be excited simultaneously, or independently, depending upon the waist diameter of the input beam.

We assume that the curvature attenuation Γ_O (dB per π radians) is proportional to that fraction of the guided energy which is beyond the critical radius, ρ_C. (Walker and Wilkinson (5)) (ρ is the radial coordinate and ρ_C is that value of ρ for which the phase velocity of the circumferentially travelling evanescent wave equals the velocity of light in the substrate medium; all energy beyond ρ_C is lost.) Thus, integrating the exponential evanescent field expression to obtain the lost energy fraction (constant amplitude-function, varying radius R):

$$\Gamma_C = \frac{K}{\xi} \exp(-2\xi C) \qquad (5)$$

where $\quad C = \rho_C - R = R.\left[\frac{n_e}{n_s} - 1\right]$

ξ is the lateral evanescent decay constant
n_s is the substrate refractive index
K is a constant

In this experiment we were concerned simply to note a cut-off radius for each of several bend-sets diffused for different times. The cut-off radius is taken as that radius for which the attenuation in traversing 180° is 3 dB. This is estimated by visual observation, 3 dB being approximately the smallest intensity change which the eye can detect. Rearranging equation (5) we obtain:

$$\frac{1}{R_C} = \frac{2\xi}{n_s} \cdot \frac{1}{\ln\left[\frac{K}{\xi \Gamma_C (R_C)}\right]} \cdot [n_e - n_s] \qquad (6)$$

Since the guide width is constant (we vary diffusion time) ξ a weak function of n_e except near cut-off where $\xi \to o$. Thus we expect $1/R_C$ to vary almost linearly with the index difference $n_e - n_s$, with some curvature near the origin. An experimental graph of $1/R_C$ versus n_e (Figure 4) for different diffusion times, shows the expected behaviour. Figure 3 , a photograph of the surface scatter from guiding bends, illustrates the estimation of R_C.

CONCLUSIONS

It has been shown that low-loss, narrow stripe waveguides (single mode) can be made using silver/sodium ion-exchange. Bends of radius much less than 300 µm with little loss can be made and also directional couplers with transfer lengths below 5 mm.

Both couplers and bends behave as predicted by theory.

REFERENCES

1. Walker, R.G., Wilkinson, C.D.W., and
    Wilkinson, J.A.H. 1981. To be published

2. Wernick, S., Pinner, R., 1972. "Surface
    Treatment of Aluminium " Vol. 1 (of 2).
    Robert Draper Ltd., Teddington, England.

3. Chartier, G., Collier, P., Guez, A.,
    Jaussaud, P., Won, Y. 1980, Appl. Opt,
    19, 1092-1095.

4. Ramaswamy, V., Standley, R.D., 1978,
    Bell Syst. Tech. Jo., 57, 2685-2693.

5. Walker, R.G., Wilkinson, C.D.W., 1981.
    To be published.

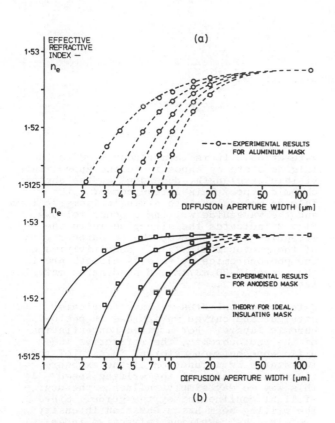

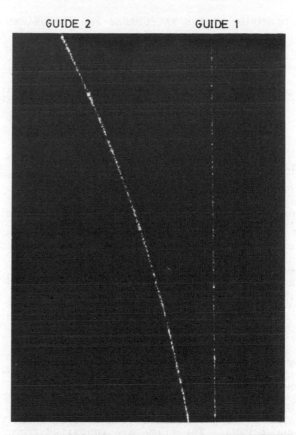

GUIDE 2   GUIDE 1

Figure 1 Experimental and theoretical
dispersion curves for stripe, ion-exchanged
waveguides (11 minutes diffusion time)

Figure 2 Surface scatter from the divergent
output arms of a directional coupler,
demonstrating coupling (I_2/I_1 = +5 dB)

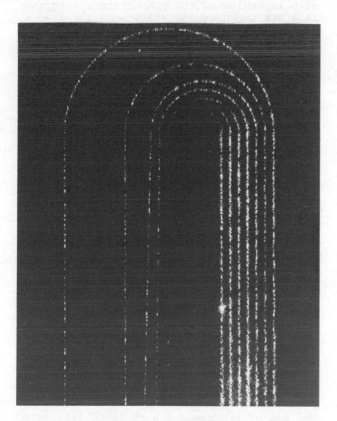

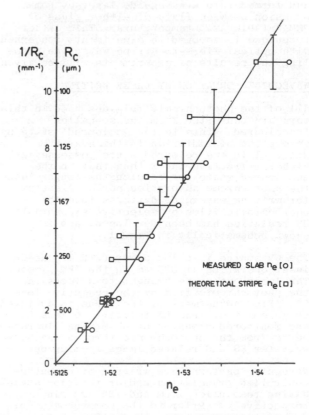

Figure 3 Surface scatter from propagating
curved waveguides, excited simultaneously.
($R_C \simeq 130$ microns)

Figure 4 Experimental graph of reciprocal cut-
off radius ($1/R_C$) against n_e for curves of
varying diffusion time.

PHOTOCHROMIC STRIPE WAVEGUIDES FOR INTEGRATED OPTICS

A G Hallam, I Bennion and W J Stewart

Plessey Research (Caswell) Limited, Allen Clark Research Centre, Caswell, Towcester, Northamptonshire NN12 8EQ

INTRODUCTION

We report the fabrication and properties of stripe optical waveguides in organic photochromic materials. The waveguides are formed by the relatively simple process of passing the planar photochromic sample through a focussed UV laser beam. Some materials from this class of photochromics have been reported by us previously, when the performance of grating components in planar waveguides was discussed (1,2). In the present paper we also report on two new materials exhibiting improved properties for longer wavelength guided propagation.

The photochromic materials are organic fulgides (3) referred to, for convenience, by a type-number, eg. type 540. Much of the present work involves photochromic samples deposited in solid solution in poly (methyl methacrylate) (PMMA) as thin films on substrates of either fused silica at commercially produced PMMA. The films are deposited from a liquid solution in 2-ethoxy ethyl acetate by the flow-coating procedure described in (1), followed by baking to render them isotropic. A second deposition technique has been used in which the photochromic is dissolved into a thermoplastic alkyl resin and formed into a waveguide layer by compression between flats of either glass or PMMA at elevated temperature. This latter technique is adopted in experiments concerned with optical fibre-to-stripe waveguide coupling and results of experiments will be given.

WAVEGUIDE FORMATION BY LASER WRITING

All of the photochromic fulgides used in this work are converted from the so-called 'uncoloured' state to the 'coloured' state by absorption of radiation in the near-UV. Material in the coloured state possesses a higher refractive index than that in the uncoloured state for wavelengths longer than the photochrome absorption peak. A system for writing waveguide patterns into the photochromic films by selective exposure to UV radiation has been developed, and is shown schematically in Fig. 1.

The UV source is a low power (2-3 mW) He-Cd laser operating at 325 nm in the TEM_{OO} mode. The laser output is brought to a focus in the photochromic film by the objective lens L3. Fine adjustment of the focus is achieved by movement of lens L1 relative to lens L2, and monitored visually by observing fluorescence from the photochromic through the beamsplitter BS and related image forming optics.

Waveguide patterns are written by computer-controlled translation and/or rotation stages having resolution 1 μm and .25 arc min, respectively. Prior to the commencement of any writing operation, the entire photochromic sample is given a uniform exposure to a low power UV lamp and coloured to

saturation. It is then reconverted to the fulgide state by exposure to the appropriate visible wavelength, dependent on the photochromic type. This action converts the fulgide molecules into a single isomeric form and the waveguide writing process follows. Fig. 2 indicates the situation where the waveguide is defined by the coloured region of the photochromic film. Sensitivities of the photochromics at this wavelength are in the range 10-30 mJ/mm^2, dependent on material type.

Using this apparatus we have fabricated straight and curved waveguides in photochromic layers. For a given concentration of the photochromic, the refractive index difference between material in each of the two states is a function of the exposure given and, hence, of the writing speed. In this way we may simply influence the degree of light confinement by the guide. Since the writing beam has a Gaussian intensity profile, the resulting refractive index profiles possess no abrupt changes and it is reasonable to expect that propagation losses induced by waveguide boundary roughness would be low. This has been confirmed experimentally by comparing the loss in a slab waveguide of the photochromic with that in a stripe waveguide written in the same sample. Losses, measured at 0.8 μm wavelength using prism-film coupling, typically do not exceed ∿1 dB/cm.

More complex waveguide patterns may be written simply by software modification. By introducing a gradual defocus of the writing beam during translation of a photochromic sample, a waveguide of tapered width may be formed. Since the exposure is effectively decreased, so also is the induced refractive index difference, tending to maintain the effective index of the guide along its length and leading, simply, to a tapered coupler for connecting waveguide sections of dissimilar widths.

Full experimental details relating to structures written in various photochromics using this system will be presented.

This same system has been utilised for the photolithographic definition of waveguide patterns in photoresist, prior to deposition of titanium and in-diffusion into lithium niobate. For this application, the final objective lens is used to image an aperture on to the photoresist to achieve the relatively sharp edge definition required. Stripe widths of 1.8 μm have been produced in this way.

MATERIALS FOR LONGER WAVELENGTH OPERATION

One aim of the present work is to fabricate components for use in conjunction with single mode optical fibres operating in the 1.3 μm and 1.55 μm wavelength spectral regions.

The various photochromics and matrix materials
are selected to avoid strong absorption peaks
in the vicinity of these wavelengths. New
photochromic materials have been developed -
designated types 865 and 871 - to be used
specifically in these longer wavelength
applications, and their absorption spectra
are shown in Fig. 3, alongside those of type
540 for comparison. Operation at 1.3 μm and
beyond permits us to produce materials with
absorption spectra extending further into
the red spectral region, since the longer
wavelength lifetimes of the photochromic
states are extremely long. The considerably
greater extinction coefficients of 865 and
871 together with the longer wavelength
photochrome absorption peaks produce greater
refractive index differences in the 1.3 -
1.55 μm range than have been exhibited by any
other fulgide previously reported. Extensive
testing has proven that the new materials
are thermally stable and possess similar
properties to those of the previous fulgides
which we have studied for integrated optics
applications.

We shall report on the experimental evaluation of components formed in these new
materials.

REFERENCES

1. Bennion, I, Rogers, B D and Stewart, W J,
    1980, Proceedings Sixth European Confer-
    ence on Optical Communication, York,
    England, 260-263.

2. Stewart, C, Stewart, W J, and Rogers, B D,
    1979, Proceedings Optical Communication
    Conference, Amsterdam, Netherlands,
    Paper 15.5.

3. Heller, H.G., "Substituted Organic Photo-
    chromic Compounds and Processes for their
    Preparation", U.K. Patent No. 1464603.

ACKNOWLEDGEMENT

This work has been carried out under contract
to British Telecom Research Department.

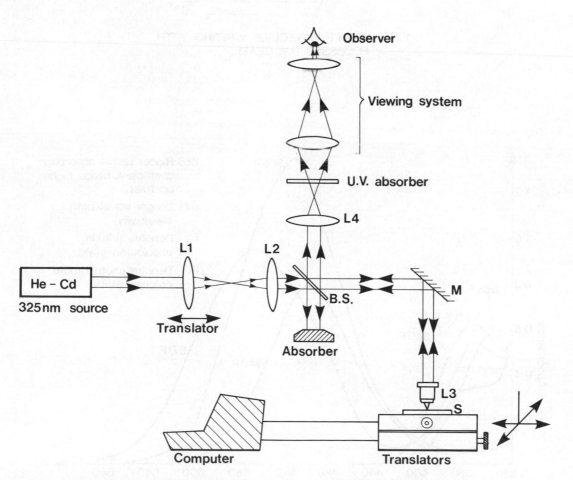

FIG.1. COMPUTER CONTROLLED WAVEGUIDE WRITING APPARATUS

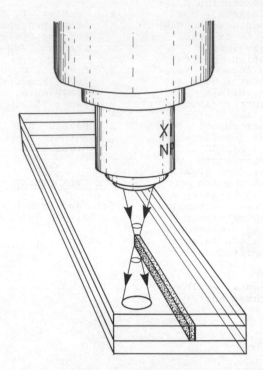

FIG.2 STRIPE WAVEGUIDE WRITING WITH
FOCUSSED U.V. BEAM

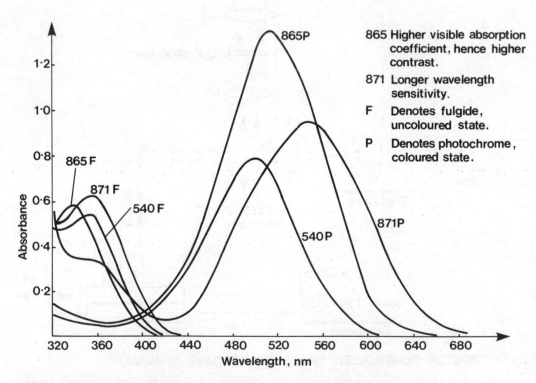

FIG.3. ABSORPTION SPECTRA OF PHOTOCHROMIC MATERIALS

SECOND ORDER EFFECTS IN GEODESIC LENSES

J. van der Donk and J. Vandewege

Laboratory of Electromagnetics and Acoustics, University of Ghent, Ghent, Belgium

INTRODUCTION

In most applications integrated optic lenses, especially geodesic lenses, are used as close as possible to the diffraction limit. In signal processing devices this is particularly important in order to achieve the desired frequency or time resolution. This means that geometric optics does not give enough information since diffraction effects are ignored. In this paper geodesic lenses are analysed by means of a beam propagation method (BPM) that fully includes diffraction. This method is first used to compute the focal fields and a comparison is made with the results obtained by means of gaussian optics. In a second part the BPM is also used to analyse the effects of fabrication errors on the quality of lenses. The materials used for signal processing are often anisotropic. This can also affect the quality of the lenses. We have analysed how important this can be.

THE BEAM PROPAGATION METHOD (BPM)

The application of the BPM to the analysis of geodesic lenses can be summarised as follows. In a first step the geodesic lens is transformed into its equivalent generalised Luneburg lens (1). The resulting refractive index profile is divided, along the direction of propagation, into small sections and each section is idealised as a thin lens placed in an homogeneous medium (2), (3). This results in a discrete set of thin lenses, through which one can propagate any arbitrary field by means of the techniques of Fourier optics. This means that the BPM includes diffraction effects. The method can be efficiently implemented by means of the Fast Fourier Transform.

FOCAL FIELDS

The BPM has been used to calculate the field in the focus of a lens with a profile proposed by Sottini et al. (4). Such a lens is described by four parameters : A and B; the radii of the circles that are imaged upon each other, C; the overall radius of the lens, D; the overall radius minus the rounded edge. Fig. 1 shows the focal fields of such a lens for gaussian input fields with different widths. The many sidelobes in fig. 1.b are due to the fact that the input gaussian spreads out outside the lens so that the finite aperture of the lens and the propagation through the rounded edge have an effect on the focal field.

Table 1 shows a comparison between calculated focal widths and the results of gaussian optics. We conclude that the calculated values are larger than the predicted ones and that the difference grows with increasing beamwidth. This is due to the fact that gaussian optics assumes infinite, thin lenses and that, for the larger beams, the finite aperture and the rounded edge cause supplementary broadening.

TABLE 1 - Comparison of $1/e$ focal widths as calculated with the BPM and with gaussian optics

collimated beam (µm)	focus (µm)	
	gaussian	BPM
100.	18.1	18.5
200.	9.1	9.5
400.	4.5	4.9
600.	3.0	4.0
800.	2.3	3.9

EFFECTS OF FABRICATION ERRORS AND SURFACE SCATTERING

The effects of fabrication errors and surface scattering have been analysed. It is important to do this with a method including diffraction since the lenses are used very close to the diffraction limit where it is difficult to analyse geometric and optical effects separately.

First the effects of deviations from the ideal depression profile are analysed. We have assumed circular sinusoidal perturbations of the depression. These perturbations affect the width of the focal field as well as its shape. Fig. 2 shows the 20 dB width of the envelope of the amplitude of the focal field as a function of the period and the amplitude of the perturbation. One sees that the maximum deviation from the ideal profile is not sufficient to define the fabrication tolerances for a geodesic lens. The period of the perturbation of the surface is at least as important. We can conclude that for this particular lens an accuracy of .5 µm has to be achieved provided that the period of the perturbation is not too large.

To model surface scattering we assumed localised scattering points. These points are assumed to be small gaussian fluctuations in refractive index. Fig. 3 shows focal fields of a lens made in a substrate with four randomly distributed scattering points. Although the width is not altered there is an important amount of scattering which will significantly increase the sidelobe level of the focal field.

THE EFFECT OF ANISOTROPY

It is well known that anisotropy can strongly affect the diffraction of a beam. It is therefore important to check the effect of anisotropy on the focussing properties of geodesic lenses.
Due to the curvature of the surface of the lens the orientation of the slab waveguide versus the optical axis differs from point to point. This means that the modal properties will not only be a function of direction but also of place.

We have assumed a step index monomode slab
waveguide (thickness = 1. μm, Δn = 1%,
λ_O = .83 m) at the surface of a geodesic lens
(A = ∞, B = 18.5 mm, C = 5. mm, D = 3.7 mm)
made in lithiumniobate (n_o=2.254, n_e=2.169).

The optical axis is orientated along the Y-
axis (fig. 4). We have calculated (5) the
minimum and maximum effective refractive index
of the mode in four different points of the
lens (a, b, c, d in fig. 4). Points b, c, d
are on the boundary of the inner lens and the
rounded edge, where the waveguide has a maximum slope. Table 2 shows the results. We conclude that, although the material is strongly
anisotropic, the effective refractive index
varies only slightly ($\Delta n_{eff} \approx$ .003). This
means that anisotropy can be neglected in this
case.

TABLE 2 - Minimum and maximum effective
refractive index of the fundamental
mode

point	minimum	maximum
a	2.2564	2.2582
b	2.2583	2.2590
c	2.2586	2.2591
d	2.2588	2.2591

CONCLUSION

We have analysed a number of second order
effects in geodesic lenses. The focal spot
size can accurately be predicted by gaussian
optics for relatively narrow beams. Fabrication errors and surface scattering turn out
to have a very important influence on the
focal field. We have shown how strongly this
can affect the resolution and the dynamic
range of an integrated optic signal processor.
At the other hand anisotropy does not have an
important influence for a normal choice of
crystal orientation.

REFERENCES

1. Kassai and Marom, "Aberration-corrected
    rounded edge geodesic lenses", JOSA,
    Vol. 69, N° 9, 1979, pp. 1242-1248.

2. Feit and Fleck, "Light propagation in
    graded-index optical fibers", Appl. Optics,
    Vol. 17, N° 24, 1978, pp. 3990-3998.

3. van der Donk and Lagasse, "Analysis of
    geodesic lenses by beam propagation
    method", Electronics Letters, Vol. 16,
    N° 8, 1980, pp. 292-294.

4. Sottini et al., "General solution of the
    problem of perfect geodesic lenses for
    integrated optics", JOSA, Vol. 69, N° 9,
    1979, pp. 1248-1254.

5. Gia Russo and Harris, "Wave propagation
    in anisotropic thin-film optical wave-
    guides", JOSA, Vol. 63, N° 2, 1973,
    pp. 138-145.

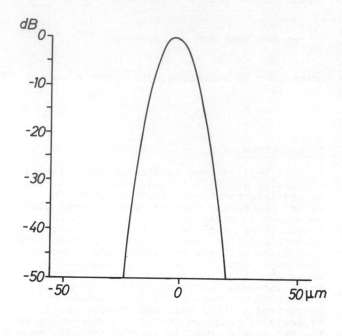

a

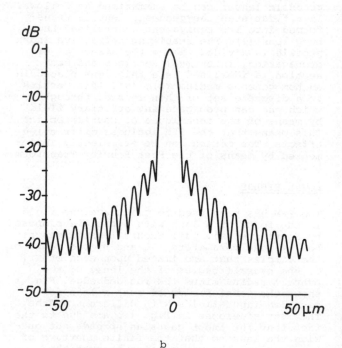

b

Fig. 1

Focal fields of a lens (A = ∞, B = 5.7 mm,
C = .8 mm, D = .65 mm) for two different
gaussian input beams : a) 200.μm b) 600.μm.

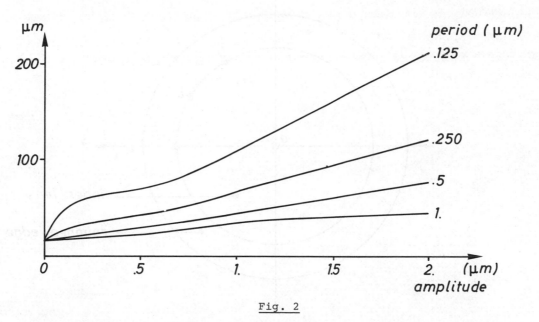

Fig. 2

20 dB width of the envelope of the amplitude of the focal field of a lens
(A = B = 8.mm, C = .5mm, D = .43mm) perturbated by a sinusoidal deviation.

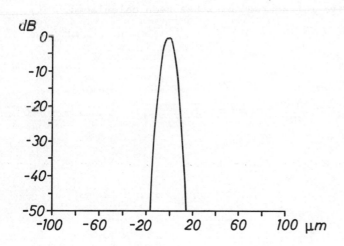

a

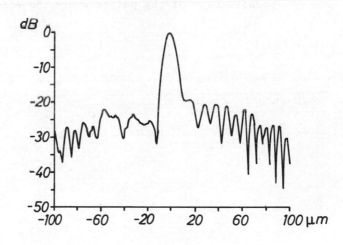

b

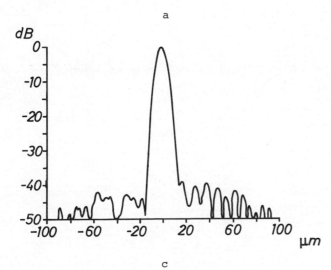

c

Fig. 3

Focal field of a lens (A = B = 8.mm, C = .5mm,
D = .43mm) in substrate showing scattering
points. a) without scattering, b) index
fluctuation = .005, c) index fluctuation =.05.

32

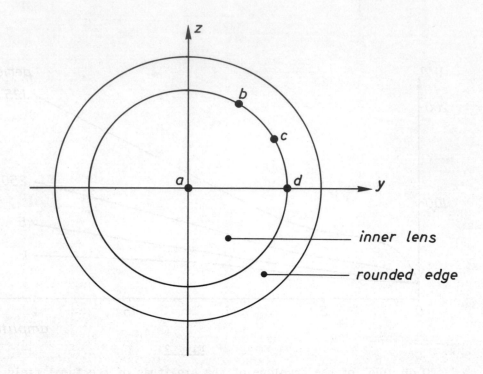

<u>Fig. 4</u>

Localisation of the points where the effective refractive index has been calculated.

SCATTERING BY PLANAR DIELECTRIC WAVEGUIDE STEP DISCONTINUITIES

G.H. Brooke

M.M.Z. Kharadly

Defence Research Establishment Pacific, Canada

University of British Columbia, Canada

INTRODUCTION

Dielectric surface waveguides, some of which involve abrupt discontinuities in cross-section, form an integral part of many complex optical waveguide systems. The proper design of such systems makes the accurate theoretical treatment of the abrupt discontinuity problem essential. To date, only Rozzi (1) has described a solution technique which is not restricted to single mode structures or to discontinuities involving small refractive index changes. Although his method has been successfully applied to TE mode scattering by a step discontinuity on a planar dielectric waveguide, it does not appear to be readily applied to the TM mode case.

In this paper, a variable-bound approach (2), in which the solution to the original (unbounded) discontinuity problem is extracted from the solution of a modified (bounded) problem, is described. The discontinuity configuration and the mode spectra in the k-plane are shown in Figure 1. Because the mode spectra of the bounded waveguide are discrete the present solution technique is mathematically less complex than previous methods, it is readily applied to either TE or TM mode cases and it is neither restricted by the physical dimensions of the discontinuity nor by the magnitude of the associated refractive index change and hence is appliable to multimode structures.

Scattering solutions, for problems comprising a wide range of discontinuities, are presented in this paper for both TE and TM modes; the TE mode results compare favourably with corresponding results obtained using Rozzi's method; the TM mode results are of comparable accuracy.

THE VARIABLE-BOUND APPROACH

The basis of the variable-bound approach lies in the one-to-one correspondence of the mode spectra of the bounded and the unbounded waveguides (3), as depicted in Figure 1, where $k_o = 2\pi/\lambda_o$ is the free-space wave number, ε_r is the relative permittivity of the dielectric and β is the propagation coefficient of a mode. Further to the one-to-one correspondence of the mode spectra, there is also correspondence of the mode fields; namely, the fast and evanescent mode fields are identical to the corresponding continuous spectrum mode fields at discrete values of β for $|x| \leq t_B$. Moreover, the slow and the surface wave mode fields, which are theoretically identical in the limit as $t_B \to \infty$, are in practice, virtually identical for t_B of the order of one or two wavelengths.

Referring to Figure 1, the variable-bound approach involves the following steps:

1. bounding the dielectric waveguide with perfect magnetic (for TE modes) and perfect electric (for TM modes) conductors,

2. solving the bounded discontinuity problem using mode-matching,

3. numerically integrating the solution of the bounded problem over a well defined range of bound locations, and

4. extracting the solution of the unbounded problem by direct mode association (i.e. surface wave scattering coefficients from the corresponding slow mode scattering coefficients).

Essentially, by bounding the open structure, the difficult original problem, involving the continuous spectrum of modes, is replaced by the relatively simple modified problem involving a discrete spectrum of modes. The latter problem can be conveniently solved, numerically, using mode-matching. The particular bound types, indicated in step 1, are chosen to avoid the low frequency cutoffs introduced by perfect electric conductors for TE modes and by perfect magnetic conductors for TM modes.

In view of the correspondence of the mode fields and of the mode spectra of the bounded and unbounded problems, an approximation to the solution of the original problem is obtained in the above procedure by going directly from step 2 to step 4. In many cases, however, the solutions of the bounded problem are sensitive to the bound location and, hence, step 3 (integration over bound locations) is required in order to achieve reasonable accuracy. This will be demonstrated by considering a numerical example involving TE mode propagation, which allows a comparison to be made with results obtained using Rozzi's method.

A Numerical Example

Referring to Figure 1, let $\varepsilon_r = 5.0$, $t_1^a = 0.07\lambda_o$, $t_1^b = 0.35\lambda_o$ and let the second TE slow mode from waveguide B be incident on the discontinuity. This mode has been chosen because its behaviour is typical of cases in which the scattering solutions of the bounded problem are sensitive to the bound location. The magnitude of the normalized reflection coefficient $|\bar{b}_2|$, of the incident mode is plotted versus t_{Bo} (i.e. t_B/λ_o) in Figure 2. The oscillating value of $|\bar{b}_2|$ appears to have the following properties: (i) a period of oscillation equal to 0.5, (ii) a constant mean value, and, (iii) a slow rate of convergence. This behaviour of $|\bar{b}_2|$ may be explained by considering the dispersion characteristics of the slow and the fast modes as functions of t_{Bo}. These characteristics, plotted in Figure 3 for waveguide B of the present example, indicate that, in strict contrast to the slow modes, the fast modes are highly dispersive functions of t_{BO}, particularly over the lower part of the range $0 \leq \beta \leq k_O$. Consequently, there is an uneven density of modes in this range. The evanescent modes (not shown here) are also highly dispersive functions of t_{BO}, however, their density is nearly uniform over the range $-j\infty \leq \beta \leq -j0$. Thus, the oscillations in $|\bar{b}_2|$ are due to the variable density of the fast modes over the lower part spectral range $0 \leq \beta \leq k_O$ and, hence, to their variable contribution to the scattering solutions. The period of oscillation is due to the transition of an evanescent mode into the fast mode spectrum at regular intervals of t_{BO} equal to 0.5. Moreover, the oscillations are, to a good approximation, about a mean value because the slow modes are only slightly and diminishingly dependent on t_{BO} and, there is slow convergence of the oscillations due to the relatively slow increase in the density of fast modes in the lower part of the fast mode spectrum.

The oscillatory behaviour of slow mode coefficients, such as that shown in Figure 2 makes the accurate solution of the original problem, using only steps 1, 2, and 4 of the solution sequence, (as discussed previously) difficult, if not impossible. This may be seen by observing the number of modes, N, required, on each side of the discontinuity, to give a certain accuracy in the mode-matching. The value of $\bar{b}_2$ is tabulated versus N and for two values of t_{BO} in Table 1. Values of P_e, the power conservation error in the numerical calculations, are also given.

TABLE 1: $\bar{b}_2$ versus N and t_{BO}

N	$t_{BO} = 2.0$		$t_{BO} = 4.0$	
	$|\bar{b}_2|$	P_e	$|\bar{b}_2|$	P_e
5	.4237	.5974	.2203	8.58
10	.4050	.0136	.4444	.2330
15	.4081	.0013	.4299	.0132
20	.4081	.0002	.4265	.0057
30	.4082	2×10^{-5}	.4294	.0009
40	.4082	1×10^{-5}	.4294	.0001

The above results indicate that as t_{BO} increases, N must also increase to achieve the same relative accuracy of the solutions. In order to achieve reasonable accuracy using only steps 1, 2, and 4, the slow convergence property of the slow mode coefficients make the required value of t_{BO}, and hence, N, so large so as to make the problem numerically impractical.

The above difficulties can be overcome by utilizing the second property of the slow mode coefficients; namely, oscillation about a mean value. Estimates of the mean value of all the slow mode coefficients, in the previous example, were calculated by averaging the solutions obtained at intervals of $\delta_{tB} = 0.1$ in t_{BO} over various periods, Δ_{tB}, of the oscillations. Hence, a crude rectangular numerical integration rule over Δ_{tB} was applied. The results are shown in Table 2. Also given in this table are the results obtained using Rozzi's method. The results in Table 2 confirm that the oscillations are, approximately about a mean value and that, by comparison with Rozzi, estimates of the mean value provide accurate approximations to the solutions of the original problem.

TABLE 2   Estimates of the mean values of the slow mode coefficients

Δ_{tB}	$|\bar{a}_1|$	$|\bar{b}_1|$	$|\bar{b}_2|$
0.9 - 1.4	.3258	.1765	.4728
1.9 - 2.4	.3262	.1795	.4790
3.9 - 4.4	.3280	.1819	.4813
Rozzi	.3272	.1814	.4820

In the above table, a_1 and b_1 are the scattering coefficients of the dominant slow modes in waveguides A and B, respectively.

Discussion of Solutions

The solution of either the original (unbounded) or the modified (bounded) discontinuity problem requires all spectral components to be accurately accounted for. In the original problem this implies that a difficult integration over the continuous spectrum be performed, whereas, in the modified problem, this implies that only a simple summation be performed. The evidence presented thus far suggests that the summation over the fast and the evanescent modes of the bounded problem is a means of performing a numerical integration over the continuous spectrum in the corresponding unbounded problem. In many cases, due to the dispersive nature of the fast modes, this numerical integration is made accurate only at great computational expense. However, the variable-bound approach, in which a further numerical integration of the solutions over a range of bound locations is performed, allows an accurate numerical integration over the continuous spectrum to be made in a computationally efficient manner. That is, by artificially increasing the coverage of the spectral range, $0 \leq \beta \leq k_O$ provided by the fast modes, the calculations may be performed with t_{BO} of the order of one or two and, hence, with N of the order of 10 or 20.

TE AND TM MODE SCATTERING

In order to illustrate the application of the integrated-bound approach, a wide range of discontinuities ($\varepsilon_r = 5.0$, $t_1^a = 0.2t_1^b$ and $0.05 = t_1^b/\lambda_O \leq 0.70$) is considered. The results for selected scattering coefficients are shown by the solid lines in Figure 4 for TE modes and Figure 5 for TM modes. These graphs are labelled in standard scattering matrix notation, S_{ij}, where subscript 1 refers to the slow mode in waveguide A and subscripts 2, 3, and 4 refer to the first three slow modes, respectively, in waveguide B. Thus, the normalized reflection coefficients for each mode have been plotted. Corresponding results for the total amount of radiation (normalized) caused by the discontinuity are shown as the solid lines labelled A_1, B_1, B_2, B_3 in each figure. These results, calculated by numerically integrating the total normalized fast mode power over Δ_{tB}, represent the radiation associated with incidence of the slow mode in waveguide A (curve A_1) and with the three slow modes in waveguide B (curves B_1, B_2, and B_3, respectively).

In Figures 4 and 5, the solid lines were generally calculated using $\delta_{tB} = 0.1$, a Δ_{tB} defined over 0.9 to 1.4, and N = 10 (TE modes) or N = 15 (TM modes). In some cases, in which the scattering coefficients had large oscillations as a function t_{BO}, $\delta_{tB} = 0.05$ was necessary. In other cases, in which the field extent of the slow modes was large, Δt_B and N were necessarily defined over larger values (i.e. in the TM mode case with $t_1^b \le 0.2\lambda_o$, Δt_B over 3.9 to 4.4, and N = 30 were used). It is interesting to note that more modes were required for the TM case. This is due to the presence of field singularities in the plane of the discontinuity in this case.

The dashed lines in Figure 4 are the corresponding results calculated using Rozzi's method. The excellent agreement between the results obtained by the two methods confirms that the variable-bound approach is effectively a means of performing a numerical integration over the continuous spectrum of the original problem and apparently as accurate a means as that proposed by Rozzi. By comparison with the TE mode case it is suggested that the results in Figure 5 for TM modes are of comparable accuracy.

CONCLUSIONS

In this paper an analytical solution technique for the problem of scattering by an abrupt discontinuity on an open planar dielectric waveguide has been presented. The method, a variable-bound approach, involves extracting the solution of the unbounded problem from a corresponding bounded problem. It was shown that, by utilizing the dispersive properties of the modes of the bounded waveguide, the complete mode spectra was simply and accurately accounted for in a numerically efficient manner. This was demonstrated by considering a wide range of discontinuity problems involving both TE and TM mode propagation. The accuracy of the results in the TE mode case was established by comparing with corresponding results obtained by Rozzi's method (1). The results for the TM mode case are believed to be of comparable accuracy, although no previous results exist for comparison.

The results presented in this paper essentially indicate that the variable-bound approach provides an accurate means of performing a numerical integration over the continuous spectrum of modes of the unbounded waveguide problem. The solution technique, therefore, depends only on the one-to-one correspondence of the bounded and unbounded mode spectra and, hence, is not in any way restricted by the physical properties of the discontinuity.

More detailed analysis, evaluation of the method presented here and interpretation of the scattering for the TE and the TM cases are given in reference (2).

REFERENCES

1. Rozzi, T.E., 1978, IEEE Trans. Microwave Theory Tech., MTT-26, 738-746.

2. Brooke, G.H., and Kharadly, M.M.Z., submitted to IEEE Trans. Microwave Theory Tech.

3. Brooke, G.H., and Kharadly, M.M.Z., 1976, Electron Lett., 12, 473-475.

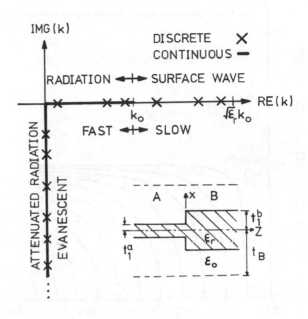

Figure 1 Discontinuity configuration and mode spectra in the k-plane.

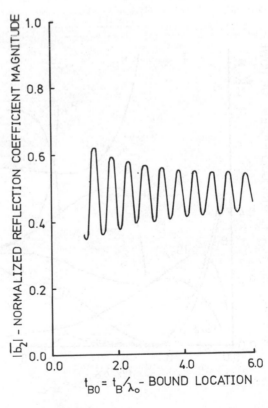

Figure 2 $|\bar{b}_2|$ against t_{Bo}: $\varepsilon_r = 5.0$, $t_1^a = 0.07\lambda_o$, $t_1^b = 0.35\lambda_o$.

36

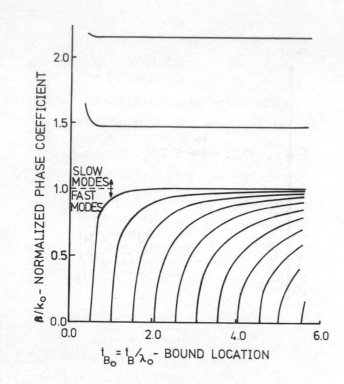

Figure 3 Modal dispersion as a function of t_{B_o}.

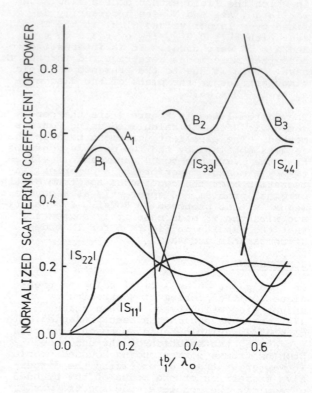

Figure 5 Scattering coefficients and radiated power against t_1^b/λ_o: TM modes, $\varepsilon_r=5.0$, $t_1^a=0.2\ t_1^b$.

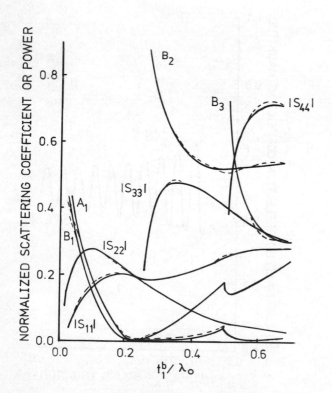

Figure 4 Scattering coefficients and radiated power against t_1^b/λ_o: TE modes, $\varepsilon_r=5.0$, $t_1^a=0.2\ t_1^b$.

TAPERED OPTICAL MONOMODE WAVEGUIDE NEAR FIELD RADIATION PATTERNS

P. Bassi, Zang Teh Yu[*] and D.B. Ostrowsky[**]

University of Bologna, Italy
[*]Semiconductor Research Institute, Shanghai, People's Republic of China
[**]Université de Nice, France

Tapered optical waveguides have been shown to offer interesting possibilities for both end-fire (Tien and Martin (1)) and evanescent (Dalgoutte et al (2)) coupling to optical fibers. Both the theoretical and experimental studies ((1) and Tien et al (3)) of their radiation patterns have been limited, however, to the far field case. Since in most coupling situations it is the near field behavior that is of interest, we have extended these studies to this region. The experimental results are in reasonably good agreement with the theoretical predictions.

A typical tapered waveguide configuration is shown in figure 1. The taper angles studied range from 0.1 to 1 deg. and the output face has been polished at an angle to permit measuring the radiation pattern at different distances from the tapers origin.

To theoretically analyze this configuration, we considered the taper to be composed of a series of discrete steps in the waveguide (figure 2).
For each of the resulting guide configurations as well as for the original waveguide, we compute the value of the propagation constant β and the spectrum of radiation modes, using the appropriate guide index profile (step or gradient). At each discontinuity we apply continuity conditions between the guided mode before and the modes (guided and radiation) after the discontinuity (Marcuse (4)). The overlap integral between the guided modes gives the percentage of power which remains guided. This fixes the total radiation power, which is coupled to the radiation modes, proportionally to the overlap integrals between each radiation mode and the guided mode of the preceeding structure.
Radiation modes have a continuous spectrum, but only a limited number can be considered : we use those necessary to describe the radiation pattern with the required spatial resolution.
Once power is transferred to a radiation mode, we assume it propagates in the substrate without any other interaction. In the section where the near field radiation pattern is calculated, the modes are summed with the appropriate phases; taking the square of the modulus of the field, we get the desired near field radiation pattern.

The planar monomode optical waveguides used in this study were fabricated by the ion-exchange process. Aluminosilicate substrates (n = 1.512) were immersed in melted KNO_3 bath ($\sim$338°C) for approximately 8 minutes. This yields monomode guides with an index profile given by :

$$n(x) = n_{sub} + \Delta n \ (1 - x/d - b \ (x/d)^2)$$

with Δn = .01, d = 3.4 μm, b = 0.5.
The tapers were formed either by partial immersion of the substrate (the meniscus at the surface of the KNO_3 bath forms a region of decreasing Δn) or by polishing. After the taper formation the output end of the guides (figure 1) were cut and polished at an angle to permit examining the radiation pattern at different distances from the taper origin.

The samples were then mounted in the apparatus shown in figure 3. Light from the He-Ne laser, appropriately polarized, is coupled into the guide via a prism coupler. An image of the output face is formed with the lens L2 and this image is scanned with a 50 micron pinhole. By deplacing the input beam in the y direction we can displace the output beam in order to vary the distance from the taper origin to the output face. In figure 4 we show the theoretically calculated near field pattern in a 0.2 deg. taper at 1 and 5 mm distances from the taper origin. This clearly illustrates the spreading of the radiation.
In figure 5 we show the results of the theoretical calculation and the measured results at a distance of 4 mm from the taper origin. The essential differences are a measured depression in the first maximum and less measured power in the higher order fringes. Possible sources of these anomalies include interferences with stray light injected by the prism coupler, radiation due to taper irregularities and effects due to radiation mode reflection from the taper surface, which we have not considered in our calculations.

Nevertheless, we believe that these results show that the calculation technique used gives a reasonable description of the radiation patterns obtained with tapered monomode optical guides and can be used to design couplers based on such structures.

ACKNOWLEDGMENTS

One of the authors (P. Bassi) wishes to acknowledge the Associazione Elettrotecnica ed Elettronica Italiana (AEI) for providing him the "Giovanni Someda" Fellowship which permitted his participation in this work.

This work was partially supported by a contract DRET 79/152.

REFERENCES

1. Tien, P.K., and Martin, R.J., 1971, Appl.Phys.Lett., 18, 398.

2. Dalgoutte, D.G., Smith, R.B., Achutaramayya, G., and Harris, J.H., 1975, Appl.Opt. 14, 1860-1865.

3. Tien, P.K., Smolinsky, G., and Martin, R.J., 1975, IEEE MTT-23, 79-85.

4. Marcuse, D., 1972, "Light Transmission Optics", Van Nostrand.

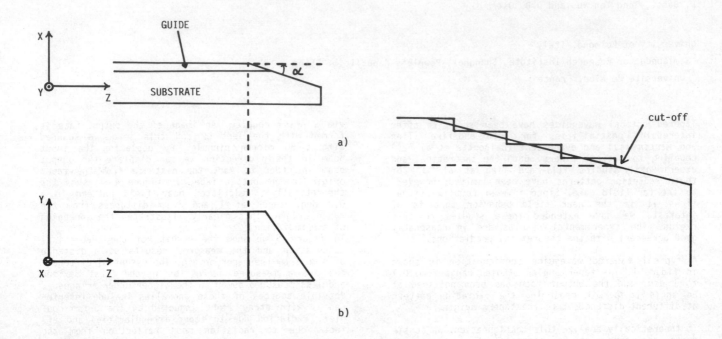

Figure 1. Tapered waveguide geometry: a) side view, b) Top view.

Figure 2. Taper model.

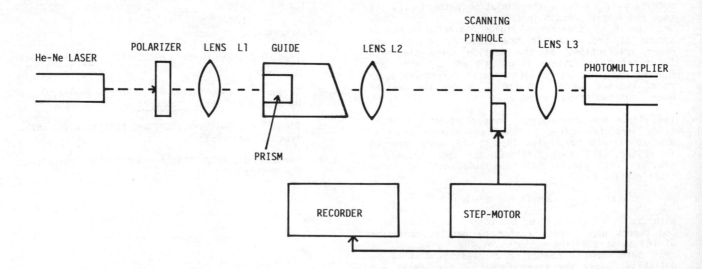

Figure 3. Experimental schematic

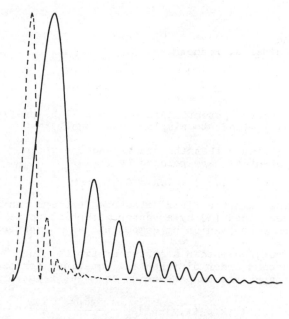

___ 25 MICRON

Figure 4. Evolution of near field pattern from 1 (dashed
line to 5 (solid line) mm from taper origin.

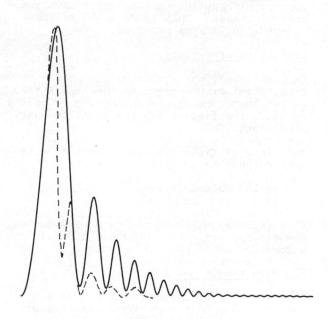

___ 25 MICRON

Figure 5. Comparison of theoretical (solid line) and
experimental (dashed line) results at 4 mm
from taper origin. The taper angle is 0.2 deg.

WAVE PROPAGATION IN AN INHOMOGENEOUS DIFFUSED CHANNEL GUIDE USING A FINITE-DIFFERENCE TECHNIQUE

J.-D. Decotignie*, O. Parriaux[0] and F.E. Gardiol*

*Ecole Polytechnique Fédérale de Lausanne, Switzerland
[0]Institut National Polytechnique, ENSIEG, St. Martin d'Hères, France

INTRODUCTION

Techniques used to fabricate optical waveguides in integrated optics, such as the diffusion of foreign species into a substrate, or ion exchange, are quite straightforward in their application. On the other hand, however, the characterisation of the propagation in the guiding structure obtained is not an easy task. As all dielectric waveguides, they are inherently open structures, in which the permittivity distribution is inhomogeneous as a result of thermal migration, while their geometry lacks of symmetry. In the planar case, the solution to the propagation problem is straightforward but it becomes quite intricate in the case of a two-dimensional transverse confinement, where all six components of the modal field must be taken into account.

As most devices in integrated optics are based on modal interactions, the knowledge of the propagation constant and of the field distribution is important. A number of approximate solutions has been proposed to the case of strip waveguides [1],[2]. In graded index structures finite element numerical techniques have been applied [3]. Though perhaps not the most elegant approach finite difference numerical techniques present the considerable advantage, in complicated structures, of solving directly Maxwell's equations and could therefore be used for checking approximate theories as well as for dealing with unsolved problems. In addition, their versatility allows one to deal with a large variety of waveguides of low mode content.

The method will be applied here to a structure obtained by in-diffusion into a semiconductor substrate. An analytical solution giving the impurity distribution was given by H.-G. Lee et al. [4]. This example should be very close to that obtained by in- or out-diffusion in dielectric substrates usually performed in integrated optics.

THEORY

Our purpose is to study the wave propagation through a channel diffused in a substrate of constant permittivity. The superstrate is air. The guide is assumed to be straight and uniform along the z direction. The permittivity of the channel is a function of the transverse coordinates x and y only. The time dependence is $\exp(j\omega t)$ and the longitudinal one is $\exp(-j\beta z)$. ω is the angular frequency and β the propagation constant. We restrict ourselves to lossless structures.

A vector potential formalism will be prefered as it leads to simpler equations.

We assume that the scalar and the vector potentials are related by :

$$j\omega\varepsilon_0\varepsilon_r(x,y)\ \Phi + \vec{\nabla}.\vec{A} = 0 \qquad (1)$$

The magnetic field is source free and can be described [5] by a vector potential whose only non vanishing components are transverse.

Then, we obtain a system of two coupled second order differential equations for the transverse components of the vector potential:

$$\nabla_t^2 A_x + (k^2\varepsilon_r(x,y)-\beta^2)A_x - \frac{\partial_x\varepsilon_r}{\varepsilon_r}\ \vec{\nabla}_t.\vec{A} = 0 \quad (2)$$

$$\nabla_t^2 A_y + (k^2\varepsilon_r(x,y)-\beta^2)A_y - \frac{\partial_y\varepsilon_r}{\varepsilon_r}\ \vec{\nabla}_t.\vec{A} = 0 \quad (3)$$

where k is the wave number in vacuum.

The differential operators are written using a finite difference formalism. A variable mesh is used in order to set the field components to zero far enough from the channel without increasing unnecessarily the number of nodes. This leads to two linear algebraic equations per node.

Conditions of symmetry

In most cases, the structure will be symmetrical with respect to the axis x = 0. We can therefore reduce the number of points by half, provided we set the right boundary conditions.

Two kinds of field configurations can exist with respect to the axis of symmetry :

a/ Ax is odd and Ay even
b/ Ax is even and Ay odd

At a point situated on the axis of symmetry, the two algebraic equations are modified according to the case of symmetry.

Boundary conditions

The permittivity distribution presents a discontinuity at the guide-air interface; therefore, equations 2 and 3 cannot be applied on the boundary and proper boundary conditions must be set. Using the continuity of the components Ez,Hz,Hx and Hy leads to :

$$A_x^- = A_x^+ \qquad ; \qquad A_y^- = A_y^+$$

$$\partial_x A_y^- - \partial_y A_x^- = \partial_x A_y^+ - \partial_y A_x^+$$

$$\varepsilon_r^+ (\partial_x A_x^- + \partial_y A_y^-) = \varepsilon_r^- (\partial_x A_x^+ + \partial_y A_y^+)$$

where indices + and − denote the values on either side of the interface.

Equations 2 and 3 are replaced by the two latter equations.

Note that if there are steps in the permittivity profile elsewhere in the structure, the same scheme applies.

All the above equations are put in a matricial form giving :

$$C \; U = 0 \qquad\qquad (4)$$

where U is a vector containing all the unknowns i.e. the values of A_x and A_y at each point of the grid. The matrix elements are functions of β.

A_x and A_y are set to zero at the outer limits of the grid. These limits can be placed as far as wanted by the use of a non regular grid. This linear system solved by searching the zeros of the determinant of matrix C.

RESULTS

Applying the technique to the case of a step index strip waveguide described by Marcatili [2] , led to an agreement within a few percents.

The technique was then applied to the case of a graded-index waveguide formed by diffusion in a homogeneous substrate. Following [4], the index distribution on the cross-section can be written as :

$$\varepsilon_r(x,y) = n^2(x,y) = \varepsilon_s(1+2\Delta f(x,y)) \qquad (5)$$

with :

$$f(x,y)=\exp(-y^2/a^2)\left[\mathrm{erf}(\tfrac{x+b}{a})-\mathrm{erf}(\tfrac{x-b}{a})\right]$$

where w = 2b is the initial width of the strip prior to diffusion, a the effective waveguide depth obtained by diffusion, $\varepsilon_s = n_s^2$ the substrate permittivity, $n_1=n(0,0)$ $\varepsilon_1 = n_1^2$ and $\Delta = (n_1 - n_s)/n_s$ the relative index difference.

Figure 1 shows a typical grid covering half of the waveguide cross-section. The grid is fine where the modal field is expected to vary significantly, and coarser far from the core.

Dispersion curves of the fundamental mode propagating in the structure have been computed in terms of a normalized frequency $V = ak \sqrt{2\Delta\varepsilon_s}$ with the scaling factor s = w/a as a parameter. The modal effective index is contained in a normalized propagation coefficient X :

$$X = \frac{k^2\varepsilon_1-\beta^2}{k^2(\varepsilon_1-\varepsilon_s)}$$

The curves are sketched in figure 2; their continuous part corresponds to the

frequency region of single mode propagation. Two distinct classes of modes propagate according to the symmetry conditions choosen at the axis x = 0. Modes sketched in figure 2 belong to those exbiting a transverse E field mainly directed towards the x-axis; they are called E^x modes [2]. The modes with electric field polarized in the y-direction, called E^y, propagate at a different phase velocity. This birefringence is shown in figure 3 by the difference between normalized propagation constants X versus V with s as a parameter. It can be noticed that the maximum of birefringence appears in the single mode region.

Once the propagation constant is found, the system (4) can be solved for the field components. Figure 4 shows the Hy components of E_{11}^x mode on a half cross section of a waveguide characterized by s = 4 and V = 3.

CONCLUSION

Finite difference numerical method presents the advantage of solving directly Maxwell's equations in their differential form. The formulation of the problem is simpler and quicker than in the case of finite elements.

It was shown that propagation in complicated structures can be readily dealt with.

REFERENCES

1. Schlosser, W., Unger, H., 1966, "Advances in microwaves", Academic Press, New-York, U.S.A.

2. Marcatili, E., 1969, Bell Sys. Tech. J., 48, 2071-2102.

3. Vandelbulke, P., Lagasse, P., 1976, Electron. Lett., 12, 120-122.

4. Lee, H.-G. et Al., 1978, J. Solid-State Circuits, SC-13, 455-461.

5. Duschek, A., 1958, "Vorlesungen über höhere mathematik" vol. 2, Springer, Vienna, Austria.

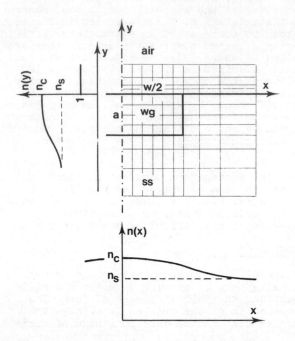

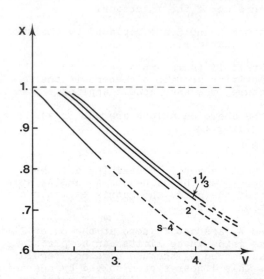

Figure 1. Waveguide wg and substrate ss cross-section of a diffused structure covered with a non regular grid. n(x), n(y), is the x,y, index dependence on axis y=0 , x=0, resp.

Figure 2. Normalized propagation constant X of E_{11}^x modes versus normalized frequency V with s=w/a as a parameter. The dashed part of the dispersion curves represent multimode propagation.

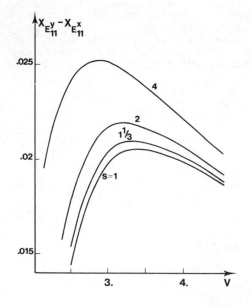

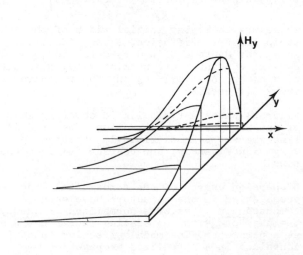

Figure 3. Birefringence ΔX between E_{11}^y and E_{11}^x modes versus normalized frequency V with s as a parameter.

Figure 4. Map of the H_y field component of the E_{11}^x mode for a value of the shape parameter s = 4, with V = 3.

OPTIMIZATION OF SELF-IMAGING PROPERTIES IN PLANAR OPTICAL WAVEGUIDES

E. Topuz

Technical University of Istanbul, Turkey

INTRODUCTION

The direct transmission of pictorial information in multi-mode optical waveguides has received considerable attention in the recent literature (1-7). Multi-mode guides have the potential of transmitting a much greater number of resolution elements (modes) in a given crossection, as compared to the conventional image guides consisting of a bundle of single-mode guides. In a multi-mode guide, an input image at $z = 0$, expressible in terms of the guided modes of the structure will be immediately blurred due to the different phase shifts incurred by different resolution elements. However, as one progresses along the guide one may expect to reach a set of points $z_p = pz_1$, $p = 1,2,..$ where all modes interact in phase coherence to reproduce an exact replica of the input image : the Fresnel image. The existence of Fresnel images requires special functional relationships between the eigenvalues of the multimode guide. Such relationships are only approximately satisfied in known realizable waveguides, thereby severely impairing their image transmission characteristics.

The purpose of the work reported here, is to demonstrate that by perturbing a multi-mode guide appropriately one may significantly improve on its image transmission characteristics. Homogeneous, symmetric optical films and a perturbation scheme based on the utilization of multi-layer homogeneous cladding will be considered in this report.

Multi-layer optical films

We consider the symmetric optical film with L homogeneous layers as shown in Fig.1, and the modes which extend to all layers. It will be assumed that the fields are independent of the y coordinate ($\partial/\partial y = 0$). The refractive indices n_ℓ of the layers will be normalized as,

$$N_\ell = \frac{n_1^2}{n_\ell^2} \quad \ell < L+1 \; ; \quad N_{L+1} > N_\ell \qquad (1)$$

The eigenvalue equation for the guided modes can be expressed as,

$$K_1 t_1 + \left[\pi/2 - g_1\right] - m\pi/2 = 0 \quad , \quad m = 1,2,...,M \qquad (2)$$

$$g_{\ell-1} = \tan^{-1}\left[\phi_{\ell-1} \tan(K_\ell t_\ell + g_\ell)\right] \; , \; \ell = 2,3,...,L \qquad (3a)$$

$$g_L = \phi_L$$

$$K_\ell = \sqrt{k_0^2 n_\ell^2 - \chi^2} \; , \; \ell \leq L \; ; \quad K_{L+1} = \sqrt{\chi^2 - k_0^2 n_{L+1}^2} \qquad (3b)$$

$$\phi_\ell = \tan^{-1} \frac{N_{\ell+1}^s \, K_{\ell+1}}{N_\ell^s \, K_\ell} \; , \; \ell \leq L+1 \qquad (3c)$$

where $k_0 = \omega \sqrt{\varepsilon_0 \mu_0}$, m is the mode index, χ_m is the axial propagation constant and s $= 0,1$ correspond to TE and TM modes, respectively. Note that, if the widths of the layers t_ℓ are adjusted to satisfy

$$t_\ell = p_\ell \pi / K_{\ell m} \quad \ell = 2,3,..L \; ; \; p_\ell = 1,2,.. \qquad (4)$$

at a given mode $\bar{m}$, then for $m = \bar{m}$, (2) will reduce to the eigenvalue equation of the unlayered film :

$$K_{1\bar{m}} t_1 + \left[\frac{\pi}{2} - g_{L\bar{m}}\right] - \bar{m} \frac{\pi}{2} = 0 \qquad (5)$$

On the other hand, the terms inside the brackets in (2) and (5) represent small contributions especially for the almost axially propagating modes. Hence, the eigenvalue equation for the multi-layered film given in (2), together with the "resonance" condition (4), can be thought of as defining a perturbation for the eigenvalues of the unlayered film given in (5), with the two sets of eigenvalues coinciding at $m = \bar{m}$. The difference between the eigenvalues of the layered and unlayered films will depend on how closely (4) is satisfied for the mode under consideration. Writing,

$$K_{\ell m} t_\ell = p_\ell \pi + \Delta_\ell(m,\bar{m}) \quad , \quad \Delta_\ell(\bar{m},\bar{m}) = 0$$
$$\ell = 2,3,...,L \qquad (6)$$

one obtains for $\Delta_\ell(m,\bar{m}) << \pi$

$$\Delta_\ell(m,\bar{m}) \cong (\pi p_\ell/2) \frac{\left[\frac{\lambda}{2t_1}\right]^2 (m^2 - \bar{m}^2)}{\frac{N_{L+1}}{N_\ell} - N_{L+1}\left[\frac{\lambda \bar{m}}{2t_1}\right]^2} \qquad (7)$$

where $\lambda = 2\pi/(k_0 n_{L+1})$. Evidently, for small values of p_ℓ and $m \approx \bar{m}$, the eigenvalues of the multi-layer film with layer widths given by (4) can indeed be considered as defining a perturbation for the eigenvalues of the unlayered film. In the following it will be shown that the remaining free parameters of the multi-layer film, i.e. the refractive indices of the layers, can be determined in such a way as to result in optimum image transmission characteristics. We begin by introducing the concept of ideal eigenvalues.

Ideal eigenvalues

The input image $\Psi_0(x)$ can be expressed in terms of the contributions $\Psi(m,x)$ of the M guided modes of the multi-mode waveguide as,

$$\Psi_0(x) = \sum_{m=1}^{M} C_m \Psi(m,x) \qquad (8)$$

where C_m are the modal excitation coefficients. Ideal imaging at $z_p = pz_1$ requires

$$\Psi(x,z) = \sum_{m=1}^{M} C_m \, \Psi(m,x) \, exp \, (-j\chi_m z) \qquad (9a)$$

$$p z_1 (\chi_{\bar{m}} - \chi_m) = 2\pi p \, P(m,\bar{m}) \qquad (9b)$$

where $P(m,\bar{m})$ is an integer function. It can be shown that the most appropriate form of P is,

$$P(m,\bar{m}) = m^2 - \bar{m}^2 \qquad (9c)$$

Hence, one can define a set of ideal eigenvalues λ_{mi} as,

$$\lambda_{mi} = \frac{\Lambda}{m \sqrt{1 - \frac{1}{N_{L+1}} \left[\frac{\lambda m}{2\Lambda}\right]^2}} \quad , \quad m = 1,2,.,\bar{M} \leq M \qquad (10a)$$

$$\Lambda(\bar{M}) = \lambda \left\{ \frac{1}{2\sqrt{N_4}} \frac{\sum\limits_{m=2}^{\bar{M}} (m^2 - 1)}{\sum\limits_{m=2}^{\bar{M}} \chi_1 - \chi_m} \right\}^{1/2} \qquad (10b)$$

where χ_m are the propagation constants of the unlayered film. Substitution of (10) into (9) reveals that the fictitious guide with the ideal eigenvalues will indeed provide ideal image transmission when observed at the Fresnel image planes at,

$$z_p = 2p \sqrt{N_{L+1}} \left(\frac{\Lambda}{\lambda}\right)^2 \qquad (11)$$

Moreover, (10b) guarantees that the ideal eigenvalues given by (10a) yield a least squares approximation for the first $\bar{M} < M$ actual eigenvalues of the unlayered film.

Optimization scheme

In the following we will regard m as a continuous variable not restricted to integer values, except when the actual eigenvalues of the waveguide are considered. A typical dependence of the eigenvalues of the unlayered film on m is depicted in Fig.2, together with the corresponding dependence of the ideal eigenvalues as defined by (10). Substituting the "cross-over" value $\bar{m}$, where $\lambda_m = \lambda_{mi}$ into (7a), one infers from (7a) and (2) that by utilizing appropriately designed layers one can perturb the eigenvalues of the unlayered film so as to obtain a better fit with the ideal eigenvalues (Fig.2), provided that the unperturbed eigenvalues satisfy,

$$\lambda_{mi} \leq \lambda_m \, , \, \bar{m} \leq m \leq \bar{M} \quad ; \quad \lambda_{mi} \geq \lambda_m \, , \, 1 \leq m \leq \bar{m} \qquad (12)$$

Highly multi-mode films wherein (12) is no longer satisfied will be excluded from the present analysis. With this restriction in mind, we will assume that the unlayered film (t_1, N_{L+1}), the permissible number of layers (L) and the number of resolution elements to be considered ($\bar{M}$) are given, and that it is desired to evaluate the widths and refractive indices of the layers so as to optimize the image transmission characteristics of the resulting layered film in some predetermined sense. One may then proceed as follows:

1- Evaluate λ_m from (5), λ_{mi} from (10) and determine $\bar{m}$.

2- Choose a small value for p_ℓ and determine the layer widths t_ℓ from (4).

3- Determine the remaining L-1 free parameters,i.e. the refractive indices of the layers, by satisfying L-1 appropriately chosen conditions.

4- Evaluate the image planes z_p from (10) and (11), using the eigenvalues of the resulting layered film.

A simple example will serve to illustrate the application of the method. We will consider TM modes in a film waveguide characterized by $t_1/\lambda = 60$ and NA = 0.15, which can transmit M = 35 modes. Choosing $\bar{M} = 25$ in defining the ideal eigenvalues one obtains $\bar{m} = 20.65$. The dependence of the phase errors due to the differences between actual and ideal propagation constants, $\chi_m - \chi_{mi}$, on the mode index is depicted in Fig.3. Next, we consider the utilization of a single layer (L = 2), and choose to design this layer to achieve an almost-equi-ripple phase error characteristic as shown in Fig.3. The additional condition in step 3 of the optimization procedure was chosen to equate one eigenvalue of the unlayered film to the corresponding ideal eigenvalue. In the illustrated case, where this eigenvalue was $m_o = 15$ (See Fig.2),the resulting parameters for $p_\ell = 1$ were $N_\ell \cong .36$ and $t_\ell/\lambda \cong .37$. For this waveguide the first Fresnel image plane is located at $z_1/\lambda \cong 4 \cdot 10^4$. If modal phase errors less than $\pi/8$ radians are considered acceptable, in accord with reference (6), one infers from Fig. 3 that in the unlayered film severe image degradation will occur beyond the second Fresnel image plane, whereas by the addition of a single layer it becomes possible to attain comparable resolution at distances larger by one order of magnitude.

Acknowledgement

The author wishes to thank to L.B.Felsen who brought this problem to his attention.

REFERENCES

1.Ulrich,R.,1975, Opt.Comm., 13, 259-264.

2.Ulrich,R.,and Ankele,G.. 1975,App.Phys.Lett. 27, 337-339.

3.Yariv,A.,1976, J.Opt.Soc.Am.,66, 301-306.

4.Gover,A.,Lee C.P.,and Yariv,A., 1976, J.Opt.Soc.Am., 66, 306-311.

5.Bryngdahl,O.,1978,J.Opt.Soc.Am.,68, 310-315.

6.Ulrich,R.,and Kamiya,T., 1978,J.Opt.Soc.Am., 68,583-592.

7.Grigor'eva,E.E.,and Semenov,A.T., 1978, Sov.J.Quant.Elec., 8, 1063-1073.

45

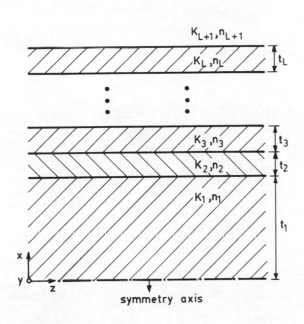

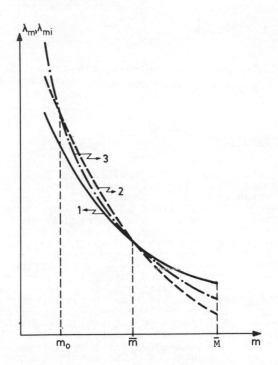

Figure 1 Multi-layered symmetric film.

Figure 2 Variation of the eigenvalues with m. (1) unperturbed, (3) perturbed, and (2) ideal eigenvalues.

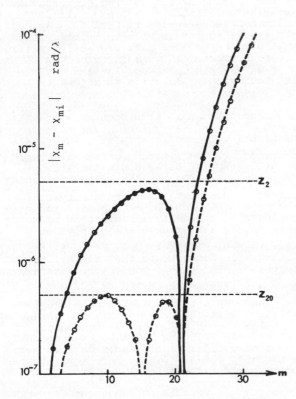

Figure 3 Variation of the phase error with m. —— : Unlayered film, ---- : The effect of one additional layer, as discussed in the text.

GUIDED-WAVE DEVICES FOR OPTICAL FIBER COMMUNICATIONS

R. C. Alferness

Bell Laboratories, Holmdel, New Jersey

INTRODUCTION

Optical communication applications have been a primary goal for guided-wave device research since its inception. The recent progress in this field together with the continued improvement and acceptance of single-mode fibers enhance the prospects for high data-rate single-mode fiber lightwave systems employing guided-wave terminal devices. In this talk we will review the recent progress toward this goal. A brief review and the current status of guided-wave switch/modulators, wavelength filters and polarization controlling devices will be discussed. Particular emphasis will be given to recent work that addresses the polarization problems in single-mode fiber systems and to high speed modulators. The latter appear particularly attractive for early incorporation into high data-rate systems. Problem areas, such as device insertion loss, and recent progress toward overcoming them will also be discussed. We will concentrate primarily on Ti diffused lithium niobate waveguide devices.

SWITCHES

Because of their versatility in both communication and signal processing applications, optical switches have received much attention throughout the evolution of guided-wave device research. Currently the reversed $\Delta\beta$ directional coupler (1, 2) and the balanced bridge switches (3, 4) have demonstrated the most promise. While switching networks for signal routing are a prospective goal for these devices, applications which take best advantage of the high speed potential are perhaps more promising. An example is high speed time division multiplexing. In addition, switches may be employed as sparing devices to allow rapid, remote replacement of failed sources or detectors with redundant elements. While early switches were effective in switching only a single linear polarization, a polarization-independent directional coupler switch that employs weighted coupling has been demonstrated (5). This switch is compatible with available single-mode fibers which do not maintain linear polarization.

MODULATORS

High speed guided-wave modulators are currently of great practical interest for use in high data-rate single-mode lightwave systems. Direct current modulation of semiconductor lasers is the most convenient means of digital signal encoding. However, wavelength instability suffered by diodes under rapid modulation may severely reduce the achieveable system bandwidth due to wavelength dispersion - especially in the very low loss, nonzero dispersion region of $\lambda \sim 1.55\mu m$. Fast encoding by external guided-wave modulators could over-

come this limitation allowing the lasers stable continuous operation.

High speed ($\gtrsim$1GHz) intensity modulation has been reported for directional coupler (6, 7, 8), Y branch interferometer (9), waveguide cutoff (10), and TE $\leftrightarrow$ TM mode converter (11) devices. Both lumped and traveling-wave electrodes (12, 13) have been demonstrated. Modulation rates approaching 10 GHz have been achieved with traveling wave modulators in spite of the velocity mismatch in LiNbO$_3$ (12). A proposed traveling-wave directional coupler gate in which very short (picosecond) optical pulses can be produced when driven by a purely sinusoidal electrical drive signal (14) offers exciting possibilities for both communication and signal processing. Several velocity matching techniques which take advantage of thin film fabrication techniques have also been proposed (15). We will discuss the relative merits of the different modulator geometries and electrode types and an overview of the current state of the art in modulators will be given.

WAVELENGTH FILTERS

Wavelength multiplexing offers enormous potential for increasing the information capacity of single-mode lightwave systems. Ti diffused lithium mobate waveguide filters are particularly appealing for wavelength multiplexing/demultiplexing because they are active devices. Thus, both electrical tunability of the filter center wavelength and a combination of the multiplexing and modulation functions in a single device are possible.

Two filter types have been demonstrated with titanium diffused waveguides. The tunable directional coupler filter yields relatively broadband ($\sim$200Å) filtering using wavelength dependent coupling between dissimilar waveguides (16). Broad tunability with low tuning voltages ($\sim$100Å/volt) has been achieved. Wavelength dependent TE $\leftrightarrow$ TM conversion using a periodic electrode to achieve phasematching at the filter center wavelength provides narrow to medium filter bandwidths (5Å to 50Å) (17, 18). To demonstrate the feasibility of integration, three mode converter filters with center wavelengths separated by 80Å and each $\sim$15Å bandwidth were fabricated on a single crystal. Crosstalk between adjacent channels was below -20 dB. Electrical tunability of mode converter filters has also been recently demonstrated (19). While early devices were demonstrated for $\lambda \simeq \cdot 6\mu m$ we have now fabricated TE $\leftrightarrow$ TM filters for $\lambda \simeq 1.3\mu m$.

To make use of wavelength dependent TE $\leftrightarrow$ TM conversion for multiplexing/demultiplexing a polarization splitter is generally required. This requirement has been recently overcome using TE $\leftrightarrow$ TM conversion between a mismatched directional coupler pair (20). Furthermore, the symmetry of the TE ↔ TM interaction results in a polarization-independent filter response.

POLARIZATION CONTROLLER

Polarization-independent devices can be employed to overcome problems resulting from polarization instability for most communication applications. However, for heterodyne systems, for example, a received signal with fixed polarization is required (21). In this case either special polarization preserving fibers or active polarization stabilization is necessary. The latter has been previously demonstrated by bulk polarization transforming elements (22). More recently the first integrated optic device capable of general polarization control has been demonstrated (23). This device combines a TE ↔ TM converter between two electro-optic phase shifters. By applying appropriate voltages to these elements, any elliptical input polarization can be transformed to any desired output polarization.

INSERTION LOSS

To date, most emphasis in guided-wave research has been given to demonstrating new devices and to optimizing speed, drive power, etc. However, for practical application of these devices low insertion loss is essential. Important work is still required to reduce fiber-waveguide and bend loss. Currently fiber-waveguide-fiber insertion loss (excluding Fresnel reflection) of ∼3 dB has been reported for ∼1 cm long straight Ti-diffused waveguide (24). Fiber coupling can be enhanced by selecting diffusion parameters to approximately match the waveguide mode to that of the fiber. However, such parameters may not be optimal for device design. It is likely that diffusion parameters which provide an acceptable trade off between device performance and insertion loss will be required. Waveguide tapers may also prove useful.

REFERENCES

1. H. Kogelnik and R. V. Schmidt, 1976, IEEE J. Quant. Electron, QE-12, 396-401.

2. R. V. Schmidt and H. Kogelnik, 1976, Appl. Phys. Lett., 29, 503-505.

3. V. Ramaswamy, M. Divino, and R. Standley, 1978, Appl. Phys. Lett., 32, 644-646.

4. M. Minakota, 1979, Appl Phys. Lett., 35, 40-42; O. Mikami and S. Zembutsu, 1979, Appl. Phys. Lett., 35, 145-147.

5. R. C. Alferness, 1979, Appl. Phys. Lett,, 35, 758-750.

6. P. S. Cross and R. V. Schmidt, 1978, IEEE J. Quant. Electron., QE-15, 1415-1518.

7. O. Mikami, J. Noda, and M. Fukuma, 1978, Trans. IECE Japan, E-61, 144-147.

8. R. C. Alferness, N. P. Economou, and L. L. Buhl, 1981, Appl. Phys. Lett., 214-216.

9. F. J. Leonberger, 1980, Opt., Lett., 5, 312-314.

10. A. Neyer and W. Sohler, 1979, Appl. Phys. Lett., 35, 256-258.

11. R. C. Alferness and L. L. Buhl, to be published.

12. M. Izutsu, Y. Yamane, and T. Sueta, 1978, IEEE J. Quant. Electron., QE-14, 394-397.

13. K. Kubota, J. Noda, and O. Mikami, 1980, IEEE J. Quant. Electron., QE-16, 754-763.

14. E. A. J. Marcatili, 1980, Appl. Opt., 19, 1468-1476.

15. For example, E. A. J. Marcatili, patent pending.

16. R. C. Alferness and R. V. Schmidt, 1978, Appl. Phys. Lett., 33, 161-163.

17. R. C. Alferness, 1980, Appl. Phys. Lett., 36, 513-515.

18. R. C. Alferness and L. L. Buhl, 1980, Opt. Lett., 5, 473-475.

19. R. C. Alferness and L. L. Buhl, this conference

20. R. C. Alferness and L. L. Buhl, 1981, Appl. Phys. Lett., 39, 131-134.

21. Y. Yamamoto and T. Kimura, 1981, IEEE J. Quant. Electron., QE-17, 919-935.

22. R. Ulrich, 1979, Appl. Phys. Lett., 35, 840-842; Y. Kidoh, Y. Suematsu, and K. Furuya, 1981, IEEE J. Quant. Electron, 991-994.

23. R. C. Alferness and L. L. Buhl, 1981, Appl. Phys. Lett., 38, 655-657.

24. M. Fukuma and J. Noda, 1980, Appl. Opt., 19, 591-597; C. H. Bulmer, S. K. Sheem, R. P. Moeller, and W. K. Burns, 1980, Appl. Phys. Letts., 37, 351-353.

AN ULTRA-WIDEBAND ACOUSTOOPTIC MODE CONVERTER USING GUIDED-WAVE STRUCTURES

Le N. Binh

Division of Forest Research, CSIRO, Australia

ABSTRACT

An ultra-wideband acoustooptic mode converter is presented. A titled-finger chirp transducer is incorporated with the confined optical and acoustic waveguide structure. The surface acoustic wave zig-zag propagation in the acoustic waveguide is discussed. The acoustic frequency and mode conversion phase matching condition is then derived. It is shown that a 500 MHz conversion bandwidth centred at 1 GHz is possible.

INTRODUCTION

The acoustooptic (AO) deflection (1) and mode conversion interactions (2) in optical integrated circuits have been extensively studied. These devices constitute a key active element in future integrated optical signal processing and communication systems. TE↔TM mode conversion is very important in integrated optics because it makes possible a number of device applications such as switches, modulators, and polarization multiplexers. Experiments on acoustooptic mode conversion have been reported in ZnO films (3) and in metal-diffused $LiNbO_3$ waveguides (4). However, optical guided waves (OGW) and surface acoustic waves (SAW), are not confined laterally in these device configurations, and high mode conversion efficiency cannot be obtained. Also very narrow conversion bandwidth (less than 1MHz) results. In polarization multiplexing techniques a wide bandwidth is a basic requirement. The experimental results (5), and theoretical approach (6), of a wideband AO mode converter using double confined structures have been reported. A bandwidth of about 30MHz has been obtained.

In this paper, we report some findings on another wideband mode converter configuration using a double-confined optical and acoustic waveguide structure, together with a tilted-finger chirp transducer.

THE WIDEBAND STRUCTURE

First we review briefly the doubly-confined structure arrangement as illustrated in Fig. 1, and its most recent development. Ti film, usually more than 1000Å thick, is sputtered and etched to a designed pattern, then diffused into $LiNbO_3$ to form fast regions for guiding SAW. A second Ti layer is then deposited and etched to a channel pattern and then diffused to form the OGW channel.

SAW propagation

We can visualize the SAW propagation in the guide by referring to Fig. 2. Region I, with a Rayleigh wave velocity v_s less than that in region II (v_f), will act as a guiding region. The guiding action is achieved by total internal reflection of the 'ray'

at the interfaces between the slow and fast regions. This propagation manner of SAW in an acoustic waveguide, is similar to a dielectric electromagnetic waveguide as pointed out by Stern (7). In addition, the percentage change in substrate parameters is quite small, and perturbation theory can be applied. We consider the substrate as a semi-infinite, elastically isotropic medium. It has been shown by Li (8) that an equivalent of an isotropic medium may be used. By this choice, the analysis is greatly simplified.

We have shown that (6) an optical analogy can be used provided the substrate is optically an isotropic medium. Thus from Snell's law in optics a critical angle can be defined as:

$$\phi_c = \cos^{-1} \frac{v_s}{v_f} \simeq \cos^{-1}(1 - \frac{\Delta v}{v})\ldots\ldots\ldots(1)$$

where

$$\Delta v = v_f - v_s << v_s, \quad v_f \text{ and } v \simeq v_s$$

using series expansion of $\cos \phi_c$ we have

$$\phi_c \simeq (2\frac{\Delta v}{v})^{1/2}$$

A similar equation, given by Coussot (9), is

$$\phi_c \simeq \cos\{\frac{\pi}{2} (p \frac{\phi}{\phi_c} - m_a)\}\ldots\ldots\ldots\ldots(2)$$

where ϕ is the reflection angle, m_a is the acoustic mode number, and $p = 4 \phi_c K_{as} a/\pi$, K_{as} is the acoustic wave vector in slow region, and 2a is the acoustic waveguide width. From eq.(2) we can obtain a relation between the acoustic frequency f_a and angle ϕ as

$$f_a = \frac{v_s}{4\phi_c ax} (2/\pi \cos^{-1}x + m_a)\ldots\ldots\ldots(3)$$

where

$$x = \phi/\phi_c$$

This dependence of f_a on reflection angle ϕ is plotted in Fig. 3(a), and an enlarged region of the curve with $m_a = 0$ (fundamental mode), is presented in Fig. 3(b). In the calculation 2a = 32.5 μm, v_s = 2769 m/s, $\Delta v/v$ = 0.015 are used. The value of 2a can be obtained through eq. (2). We observe from these curves that the waveguide can support a SAW operating at frequencies up to a few GHz. The reflection angle is proportional to $1/\phi$ and to $\cos^{-1}x$, thus f_c changes rapidly as $\phi \to 0$.

The phase matching conditions

The phase matching condition of collinear AO TE↔TM mode conversion is written as:

$$f_{ap} = \frac{v_p}{\lambda_o} (n_E - n_M)\ldots\ldots\ldots\ldots\ldots(4)$$

where f_{ap} is the acoustic frequency for

phase matching, λ_O is the optical wavelength in vacuum, and n_E, n_M are the effective indices of TE and TM OGW modes. Since the refractive index changes Δn_o and Δn_e by the ordinary in-diffusion technique are usually small, n_E and n_M can take values of LiNbO$_3$ substrates. It then can be shown that the TE$\leftrightarrow$TM mode conversion occurs with an acoustic frequency of about 500MHz and around this frequency region with other OGW modes if the optical channel could support more than one mode.

With the zig-zag manner of SAW as shown in Fig. 2, the acoustic wave vector along the x direction (i.e. the direction of OGW) is given by:

$$|\underset{\sim}{K}_{pm}| = |\underset{\sim}{K}_a| \cos \phi \quad \dots\dots\dots\dots\dots\dots(5)$$

where $\underset{\sim}{K}_a$ is the acoustic wave vector along the propagation 'ray' direction.

From eq. (5) together with Figs.3 (a) and (b), we can conclude that the bandwidth of the transducer is approximately equal to the bandwidth of an AO mode converter. We have demonstrated experimentally this fact (5).

The tilted-finger chirp transducer

The tilted finger chirp transducer is described in the work reported by Lee et al (10) and is evolved from their previous wideband deflectors. The bandwidth of this composite transducer, as with the conventional chirp transducer, should be relatively large. The transducer consists of a number of electrodes of varying periodicity, connected in parallel with each other. These electrodes are tilted at appropriate angles as shown in Fig. 4. The transducer operates at a centre frequency f_{ao} and lower and upper limits f_{al} and f_{ah} respectively.

The tilted angles ϕ and ϕ_ℓ must satisfy the conditions that

$$\phi_h, \phi_\ell \leq \phi_c \dots\dots\dots\dots\dots\dots\dots\dots(6)$$

With a transducer centred at 500MHz the synchronous frequency of the transducer can vary linearly from 320MHz to 630MHz when the finger width varies from 2.4 μm to 1.2 μm to SAW propagating in the x-direction on y-cut LiNbO$_3$ substrate (11). The diffraction of SAW is eliminated due to the acoustic waveguide. This confined structure would enable a transducer centred at 1GHz to have a bandwidth of 500MHz.

CONCLUSION

In summary, as part of an effort to devise a wideband AO mode converter configuration, we have outlined and designed a potential ultrawideband AO mode converter. A conversion bandwidth of 500MHz can be obtained. This is 500 times that of the conventional planar AO mode converter.

However it has been shown (5, 6) that the diffusion depth of the optical channel must be at minimum so that the reflection of SAW at these interfaces does not affect the normal SAW zig-zag propagation. The analyzed wideband AO mode converter reported here would certainly provide an ultra-wideband optical polarisation multiplexer in future integrated optical circuits.

REFERENCES

1. Schmidt, R.V., 1976 'Acousto-optic interactions between guided optical waves and surface acoustic waves', IEEE Trans. Son. & Utrason., SU-23, 22-32.

2. Binh, L.N., and Livingstone, J., 1980, IEE Proc., Vol. 127, Pt H., No. 6, Dec. 1980, 323-329.

3. Sasaki, et al., 1974, App. Phys. Lett., 25, 476.

4. Ohmachi Y., and Noda, J., 1977, IEEE J. Quant. Elec., QE-13, 43.

5. Binh, L.N., and Livingstone, J. 1980, A Wideband Acoustooptic TE-TM Mode Converter. IEEE J. Quant. Elect., Vol. QU-16, No. 9, 964-971.

6. Binh, L.N., (in prep.), Analysis of Acoustooptic Mode Conversion utilising Doubly Confined Optical and Acoustic Structure.

7. Stern, E., 1969, Microsound Components, Circuits and Applications, IEEE Trans Micro. Theory and Tech. (Special issue on Microwave Acoustics), MTT-17, 835-844.

8. Li, R.C.W., 1972, Proceedings of the Ultrasonic Symposium, Boston, Mass, pp263.

9. Coussot, G., 1973. Rayleigh Waveguidance on LiNbO$_3$, App. Phys. Lett. 22, (9), May 1973.

10. Lee, C.C., Liao, K.K., Change, C.L., and Tsai, C.S., 1979, Wideband Guided Wave Acoustooptic Bragg Deflector using a Tilted-Finger Chirp Transducer. IEEE J. Quant. Elect., QE-15, 10, November 1979.

11. Tanwell, R.H., 1971. Proc. IEEE, 59.

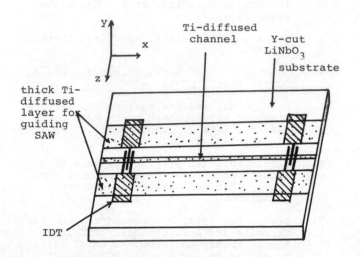

FIG.1 Basic configuration of a wideband
AO mode converter utilizing doubly
confined structure.

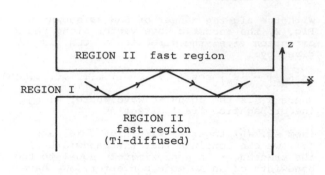

FIG. 2 Geometry of SAW propagation in
acoustic waveguide.

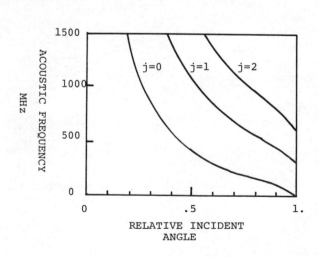

FIG.3(a) Acoustic frequency plotted vs
relative incident angle of
acoustic ray with respect to
propagation angle.

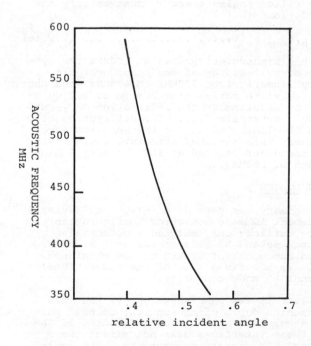

FIG.3(b) Acoustic frequency vs normalised
incident angle in diffused
acoustic waveguide, an enlarged
region around f_a = 500 MHz.

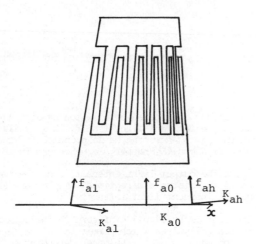 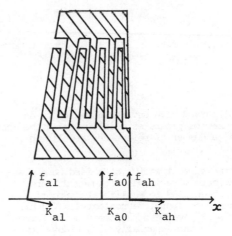

FIG.4 Tilted-finger chirp transducer used
in ultra-wideband AO mode converter.

ELECTROOPTIC INTEGRATED OPTICS

M. Papuchon

Thomson-CSF, L.C.R. - B.P. N° 10 - 91401 Orsay (France)

Electrooptical guided wave devices have led to an impressive number of papers especially during the last several years a number of references are listed in Alferness (1).

In this talk the current status of guided wave electrooptical devices is reviewed. Although different materials have been used to realize electrooptical waveguides (semiconductors, ferroelectrics...) most of the recent work have been performed using ferroelectric materials and especially with $LiNbO_3$ crystals. In fact, this is due to its excellent optical properties and to the high electrooptical coefficients which can be used when realizing guided wave devices.

The Ti in diffused technique is now widely used to create the waveguides and it has been demonstrated that the outdiffusion problem can be overcome (Ranganath and Wang (2), Chen and Pastor (3))in particular by performing the diffusion under humid atmosphere (Jackel, Ramaswamy, Lyman, Glass and Olson (4)) thus permitting the use of the waveguide with the two polarizations(TE and TM). If buffer layers are added between the guide and the electrode metal film, the losses are minimized and a true polarization insensitive device can be built. An example of this kind of circuit is shown on Fig. 1,(Alferness and Buhl (5)) where the schematic of a directional coupler filter is drawn.

This circuit which has been reported recently shows a polarization independent operation. The principle is based on the coupling between TE and TM waves in the two waveguides of a directional coupler via a periodic electrooptically induced coupling. A peak filter efficiency of 75% for a 5 Å bandwith have been demonstrated with this very promising device. With the electrooptical effect very high speed optical modulator can be realized.

In fact, the bandwith of such devices is limited by several factors which are mainly the electrode capacitance in the case of lumped type modulators and velocity mismatch for the travelling wave ones ; the later giving already very large bandwiths. Using travelling wave interaction a very high speed device has been recently reported(Sueta and Izutzu (6)).It is based on the use of an electrooptically controlled inter ferometer (Fig.2) with asymmetric planar strip electrodes. A bandwidth of 11.2 GHz has been measured on the device which permits 88% modulator with 130 mW incident electrical power. In this case only TM polarization was used. Further increase of the modulation bandwith would required velocity matching between the guided light and the microwave modulation signal and some possible schemes to achieve this have been recently compared,(Lin (7)).When using the electrooptical effect to realize modulators or switches it is very easy to study related bistable optical components.

These optical elements, simply obtained by inserting the electrooptical modulator or switch in a feedback loop can exhibit very interesting behaviour like : optical memory, differential amplification, optical clipping (see for example 8). In particular, when optical directional couplers are used, very promising devices can be obtained. For example the state of the switch can be controlled by a switching signal contained in the incident light itself. This is illustrated in Fig. 3 where the basic principle of such a device is shown,(Schnapper, Papuchon and Puech (9)).

On Fig.3a the output light intensities of the two guides of the bistable directional coupler are shown versus the incident light intensity.

The incident light signal of figure 3b includes the remote control pulses which permit to switch the light from one output to the other depending on their amplitudes relative to Iinf and Isup. The result of these switching pulses is shown on Fig.3c where the output of the two waveguides is represented versus time.

There is no doubt that high performances electrooptical devices will find applications shortly for example in the fields of : single mode fiber communications, sensors or data processing systems.

REFERENCES

1. Alferness R.C., 1981, IEEE J. Quantum Electron. Q-E17 946-959

2. Ranganath T.R. and Wang D., 1977, Appl. Phys. Lett., 30, 376-379

3. Chen B. and Pastor A.C., 1977, Appl. Phys. Lett. 30, 570-571

4. Jackel J.L., Ramaswamy R., Lyman S.P., Glass A.M. and Olson D.H., "Elimination of out diffused surface guiding in Ti diffused $LiNbO_3$"IOO C 81 Technical Digest, San Francisco, USA

5. Alferness R.C. and Buhl L.L., 1981 "Polarization-independent optical filter using interwaveguide TE ⇄ TM conversion" IOO C 81 Technical Digest, San Francisco, USA

6. Sueta T. and Izutzu M., 1981 "High speed guided wave optical components", IOO C 81 Technical Digest, San Francisco, USA

7. Lin P.L., 1981 J. Opt.Commun., 2, 2-6

8. March, 1981, IEEE J. Quantum Electronic, QE 17

9. Schnapper A., Papuchon M. and Puech C., 1981, IEEE J. Quantum Electronics, QE 17, 332-336

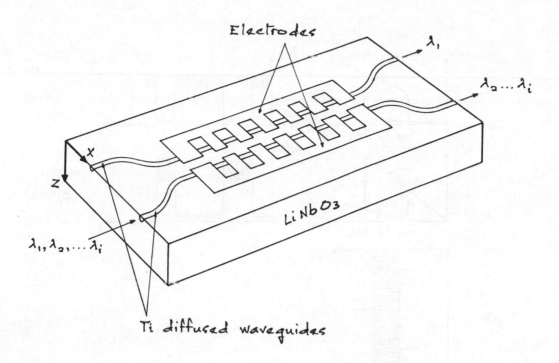

FIGURE 1 - Directional coupler filter

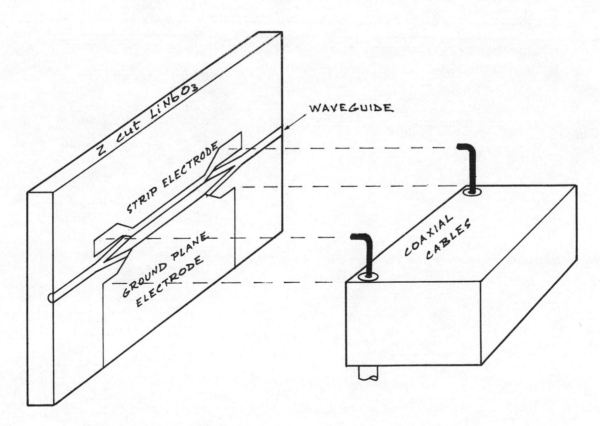

FIGURE 2 - Travelling wave amplitude modulator

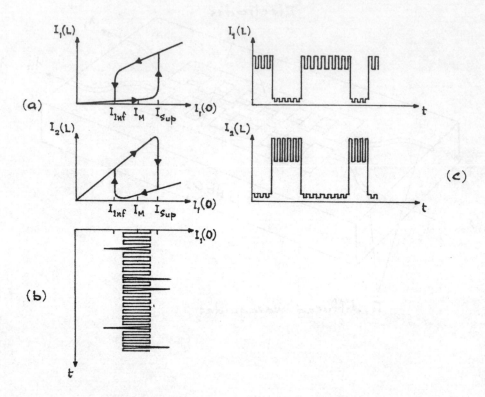

FIGURE 3 - Remotely controlled switch

TUNABLE ELECTROOPTIC TE ↔ TM CONVERTER/WAVELENGTH FILTER

R.C. Alferness and L.L. Buhl

Bell Laboratories, Holmdel, N.J.

INTRODUCTION

Integrated-optic wavelength filters are important components for wavelength multi- and demultiplexing in single-mode lightwave systems. Electrical tunability of these devices is especially desirable. For example, the oscillation wavelength of semiconductor lasers generally varies somewhat from device to device even under nominally identical fabrication conditions and also depends upon temperature and age (1). Tunable multiplexer/demultiplexers can compensate for such source wavelength changes. Recently, we reported wavelength filters based upon waveguide electrooptic TE ↔ TM conversion in Ti-diffused lithium niobate waveguides (2,3). Here we report the first demonstration of electrical tuning of these devices.

PRINCIPLES OF OPERATION

The tunable electrooptic TE ↔ TM mode converter/filter is shown schematically in Figure 1. A Ti-diffused waveguide, which supports a single TE and TM mode, is fabricated on a z cut y propagating lithium niobate crystal. In addition to the ground electrode, there are two electrodes on which voltage can be independently applied. Voltage, V_1, applied to the finger electrode provides a periodic electric field within the waveguide. The x component of this field couples to the off-diagonal r_{51} coefficient to provide TE ↔ TM conversion (2). Periodic coupling is required to achieve effective phasematching between TE and TM modes which are nonsynchronous because of the birefringence of lithium niobate. The phasematch wavelength, λ_o, is given by

$$\frac{2\pi}{\lambda_o} \left| N_{TE} - N_{TM} \right| = \frac{2\pi}{\Lambda} \qquad (1)$$

where Λ is the electrode period and N are the effective indices for the TE and TM modes. The TE ↔ TM coupling efficiency at λ_o is controlled by V_1. The phasematch condition is not satisfied for $\lambda \neq \lambda_o$ resulting in a wavelength dependent mode conversion response with filter bandwidth (FWHM), $\Delta\lambda_{BW} \approx \lambda_o\Lambda/L$ (2).

Tuning of the filter center wavelength is achieved by electrooptically changing the birefringence by applying voltage, V_2, to the uniform electrode (Figure 1). The z field component changes the TE and TM effective indices to,

$$N_{TE}(V_2) = N_{TE}(0) - \frac{\alpha_{TE}n_o^2 r_{13} V_2}{2d} \quad , \qquad (2a)$$

$$N_{TM}(V_2) = N_{TM}(0) - \frac{\alpha_{TM}n_e^2 r_{33} V_2}{2d} \quad , \qquad (2b)$$

where d is the interelectrode gap and α_{TE} and α_{TM}, the overlap integrals for the TE and TM modes respectively. Writing the voltage-tuned center wavelength as $\lambda(V_2) = \lambda(o) + \Delta\lambda(V_2)$, we find

$$\Delta\lambda(V_2) = \Lambda \; \Delta N(V_2) \quad , \qquad (3)$$

where

$$\Delta N(V_2) = - \frac{V_2}{2d} \left[n_e^3 r_{33}\alpha_{TM} - n_o^3 r_{13}\alpha_{TE} \right] \qquad (4)$$

is the electrooptically induced birefringence change. For lithium niobate $n_e^3 r_{33} - n_o^3 r_{13} \approx 2\times10^{-4} \mu m/volt$.

FABRICATION AND OPERATION

The device was fabricated as follows. A 2 μm wide, 300Å thick strip of Ti was indiffused for four hours at 980°C (4). The diffusion was performed in a water vapor rich atmosphere achieved by bubbling the argon flow gas through water to eliminate surface guiding (3). A ~1000Å thick SiO_2 layer was CV deposited on the substrate to reduce electrode loading loss. Finally the electrodes were aligned over the waveguide such that both the uniform and periodic fields affect the waveguide (Figure 1). The electrode period is, $\Lambda = 7$ μm, and the gaps between the ground electrode and the finger and uniform electrode are each ~2.5 μm. The electrode length is 3 mm. A photomicrograph of a section of the device is shown in Figure 2.

The wavelength dependent TE ↔ TM conversion efficiency, as measured with a dye laser, is shown in Figure 3. For $V_2 = 0$, nearly complete conversion is achieved for $\lambda_o \simeq 0.6$ μm for a mode converter voltage, $V_1 = 30$ volts. For a tuning voltage, $V_2 = 20$ volts the filter center wavelength is shifted by ~16Å as also shown in Figure 3. To achieve maximum conversion efficiency it was necessary to also decrease V_1 by ~3 volts which probably indicates some interaction between the two applied fields. $\Delta\lambda(V_2)$ was found to vary approximately linearly with V_2,

as predicted by Eq. (3), at a rate of
~0.75Å/volt. This corresponds to ~10 volts/
bandwidth. The measured results are in good
agreement with theory if we assume
$\alpha_{TE} \approx \alpha_{TM} \approx 0.25$. Because the electrically
induced wavelength shift scales with the
electrode period, a larger tuning rate should
be achievable for λ_o in the infrared or for
a crystal with lower birefringence such as
lithium tantalate.

In summary, using a new dual electrode
configuration we have electrically tuned the
phasematch wavelength of electrooptic
TE ↔ TM mode converter/wavelength filters.
A 10 volt tuning voltage shifts the filter
center wavelength by an amount equal to the
filter bandwidth.

REFERENCES

1. W.J. Tomlinson, 1977, Appl. Opt., 16,
   2180-2190.

2. R.C. Alferness, 1980, Appl. Phys. Lett.,
   36, 513-515.

3. R.C. Alferness and L.L. Buhl, 1980,
   Opt. Lett., 5, 473-475.

4. R.V. Schmidt and I.P. Kaminow, 1974,
   Appl. Phys. Lett., 25, 458-460.

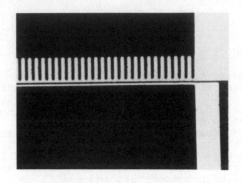

Figure 2 – Photomicrograph of a section of
the device.

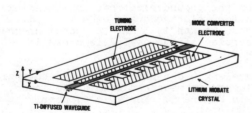

Figure 1 – Schematic drawing of the tunable
electrooptic TE ↔ TM converter/
wavelength filter.

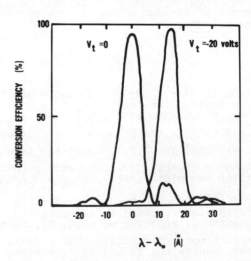

Figure 3 – Measured wavelength response with
and without application of a
tuning voltage.

EXPERIMENTS ON THE LINEARIZATION OF THE RESPONSE OF AN AMPLITUDE MODULATOR

A. Schnapper, M. Papuchon, Y. Bourbin and C. Puech

Thomson-CSF - Central Research Laboratory

For several years increasing interest has been shown in the study of bulk or integrated bistable optical devices.

They have been realized with various basic systems starting from Fabry-Perot interferometer (1) to classical light modulators (2,3).

In this paper, we will present the results obtained on the linearization of the response of an integrated Mach-Zehnder amplitude modulator (4) using an optoelectronic feedback loop.

The caracteristic response of this type of device can be expressed as

$$I_{out} \propto I_{IN}(1 + M \cos \phi)$$

Where ϕ is the phase shift difference between the two arms (which can be introduced via the electrooptical effect) and M is the modulation efficiency.

If ϕ is of the form

$$\phi = \phi_0 + A \sin \omega t$$

I_{out} can be rewritten as

$$I_{out} \propto I_{IN}\left(1 + 2M \sum_{p=0}^{\infty} J_{2p+1}\left(\frac{\pi V_M}{V_\pi}\right) \sin(2p+1)\omega t\right)$$

Where ϕ_0 is a bias phase shift (DC voltage on the electrodes) used to tune the working point of the device.

When the modulator is inserted in a feedback loop, I_{out} becomes :

$$I_{out} \propto I_{IN}\left[1 + M\left\{\left[2\sum_{p=0}^{\infty} J_{2p+1}\left(\frac{\pi V_M}{V_\pi}\right) \sin(2p+1)\omega t\right]\cos\frac{\pi G}{V_\pi}\left(\frac{kV_{IN}}{2}-V_{out}\right)\right.\right.$$
$$\left.\left. -\left[2\sum_{p=0}^{\infty} J_{2p}\left(\frac{\pi V_M}{V_\pi}\right) \cos 2p\,\omega t\right] \sin\frac{\pi G}{V_\pi}\left(\frac{kV_{IN}}{2}-V_{out}\right)\right\}\right]$$

The modulator used on the experiments has been realized using the standard diffusion technics (5), followed by the superposition of gold electrodes and the polishing of the edges of the substrate to permits end fire coupling.

The measured command voltage (V_π) of each system of electrodes is close to 8 volts for T.E. polarisation.

The distorsion of the system has been measured using the basic experimental set up shown in fig. 1.

Before amplification, a part of the output signal coming from the photodetector is applied to the input of a distorsiometer (HP 339) which permits the measurements of distorsion for different values of the amplifier gain G and the modulation voltage amplitude V_M .

The photographs of the fig. 2 show the output of the modulator for different values of the gain G of the amplifier.

The results of the distorsion measurements are shown in fig. 3 where the disorsion has been plotted versus the modulation voltage amplitude (V_M) for different gain G of the amplifier for $\phi_0 = \pi/2$. the curve G = 0 corresponds obviously to the response of the modulator without feedback and the corresponding theoritical curve is also plotted on the same diagram.

REFERENCES

1. SMITH, P.W., and TURNER, E.H., 1978 "Electrooptic non-linear Fabry-Perot devices". IEEE J. of Quantum Electronics QE 14 n° 3, 207-212

2. GARMIRE, E., ALLEN, S.D., and MARBURGER, J., 1978, "Multimode integrated optical bistable switch". Optics Letters, Vol 3 n° 2, 69-71

3. OKADA, M., and TAKIZAWA, K., 1979, "Multifunctional Electrooptic device with feedback". Japanese Journal of Applied Physics, Vol 18 n°1, 133-137

4. SCHNAPPER, A., PAPUCHON,M., and PUECH,C., 1979, "Optical bistability using an integrated two arms interferometer" Optics Com., Vol 29 n° 3, 364-368

5. HAMMER, J.M., and PHILLIPS, W., 1974, Appl. Phys. Letters 24, 545

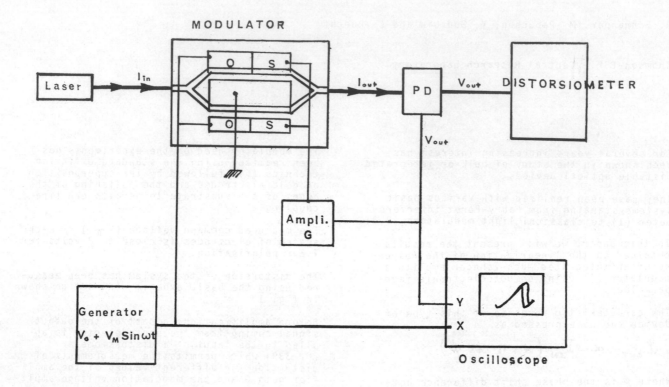

Fig. 1 Experimental set up

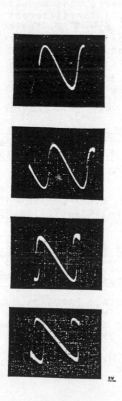

Fig. 2 Output intensity versus Applied
modulation G = 100, 200, 500 and 1000

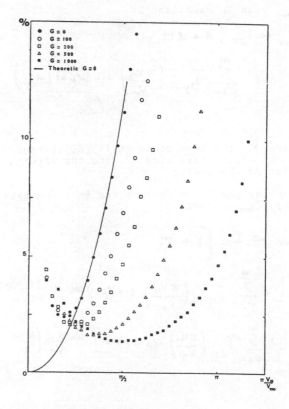

Fig. 3 Distorsiometer measurement

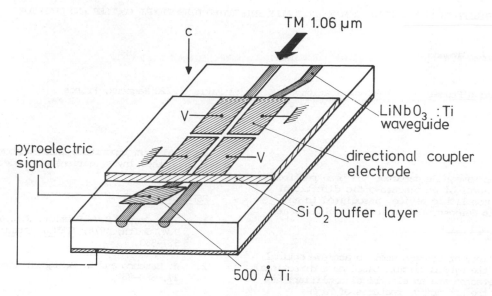

TM 1.06 µm

c

V

V

LiNbO$_3$:Ti
waveguide

pyroelectric
signal

directional coupler
electrode

Si O$_2$ buffer layer

500 Å Ti

Figure 1 Schematic diagram of the integration of a 9mm long directional coupler with a 1mm long pyroelectric
detector in LiNbO$_3$.

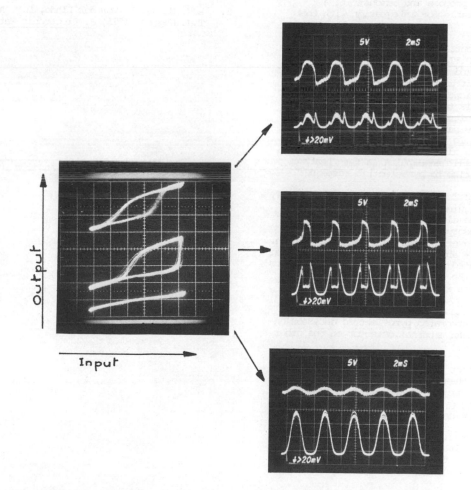

Figure 2 Bistable operation at 1.06µm of the LiNbO$_3$ integrated circuit.
Left : hysteresis loops of output versus input light power ; the curve shape is changed by varying the amplifier dc
offset voltage (-8, -11, -12V from top).
Right : display function of time of the corresponding output light detected on both channels of the directional
coupler.
upper trace : pyroelectric signal.
lower trace : external detector signal.

OPTICAL NETWORK ANALYSIS BY PHASE MODULATED SINGLE SIDEBAND DETECTION

E. Voges, O. Ostwald*, B. Schiek*, A. Neyer

FernUniversität Hagen, *Ruhr-Universität Bochum, Germany

INTRODUCTION

The amplitude and phase of optical signals
can linearly be transposed into the amplitude
and phase of a low frequency signal by single
sideband homodyne detection, where the single
sideband generation is achieved with an arbitrary odd phase modulating function of appropriate peak phase deviation. This technique
for independent amplitude and phase measurements (network analysis) is of particular
importance at optical frequencies as an replacement of the difficult heterodyne methods
and because high performance integrated-optic
phase modulators are available. Phase modulated single sideband detection is well known
at microwave frequencies for special phase
modulating functions /1,2/, and has recently
/3/ been generalized to arbitrary phase modulating functions.

BASIC PHASE MODULATION SYSTEM

The basic interferometer arrangement of fig.
1a includes in one channel the test device of
amplitude transmission A and phase shift Ψ
and a phasemodulation $\phi(\omega t)$ for single sideband generation. The bandpass (BP) filtered
signal of the photo diode

$$i_1(\omega t) \sim A(1 + \varepsilon_A) \; \cos(\omega t + \Psi + \varepsilon_p) \qquad (1)$$

contains amplitude and phase errors ε_A, ε_p
whose maximum values ε_{Amax}, ε_{Pmax} depend on
the suppression s of the disturbing image
sideband as given in eq. (2)

$$s = \int_0^\pi G(x)\cos(\phi(x)+x)\,dx / \int_0^\pi G(x)\cos(\phi(x)-x)\,dx \qquad (2)$$

for an odd phase modulation $\phi(\omega t)$, and an
additional even amplitude modulation $G(\omega t)$.
For s < 1 (the lower sideband is reduced) we
have

$$\varepsilon_{A\,max} = s \;, \qquad \varepsilon_{P\,max} = \sin^{-1}s \qquad (3)$$

and the + sign in eq. (1) applies. For s > 1
s has to be replaced by 1/s in eq. (3), and
the − sign in eq. (1) is valid.

For odd phase modulating functions the image
sideband suppression only depends on the peak
phase deviation $\Delta\phi$, the details of the modulating function are not important. A complete
suppression (s = 0 or s → ∞) is always achieved
for certain values of $\Delta\phi$.

The introduction of an additional even amplitude modulation $G(\omega t)$ preferably into the low
frequency path (fig. 1b) after blocking and
restoring the dc-level of $\cos(\phi(\omega t) + \Psi)$
allows for low amplitude and phase errors over
a wider range of $\Delta\phi$ as shown in fig. 2 for
a sawtooth phase modulation and a triangular
amplitude modulation.

For s < −40 dB we have absolute amplitude and
phase errors less than 1% and 1° respectively.

EXPERIMENTAL RESULTS

This new method and the theoretical results
are confirmed by test experiments with integrated Mach-Zehnder interferometers on LiNbO$_3$
(fig. 3a), where a phase shift Ψ to be measured
is supplied by a bias voltage U$_\Psi$, and where
the phase modulation $\phi(\omega t)$ is provided by a
sawtooth voltage U$_\phi$. The half-wave voltage
for TM-polarization is U$_\pi$ = 5.4V. The amplitude
modulation is performed as shown in fig. 1b
by using a 1495 analogue multiplier, and the
dc-restoring is achieved by clamping for
$\Delta\phi > \pi/2$. Fig. 3b demonstrates the basic technique for $G(\omega t) = 1$. The sawtooth phase modulation at 1 kHz with $\theta = 0.7\pi$ is shown in the
upper trace. The middle traces show the diode
signals $i_1(\omega t)$ for 2.7V difference of U$_\Psi$, the
induced phase shift is shown by the lower
trace of the filtered signals $i_1(\omega t)$. A comparison of the theoretical and experimental
image sideband suppression in dependence on
the peak phase deviation is shown in fig. 4
for a 1kHz sawtooth phase modulation with
$\theta = 0.9\pi$ (solid curve), and for an additional
amplitude modulation with g = 1 (dashed curve).
The experimental points (● ○) are determined
from the ratio (s+1)(s−1) of the maximum and
minimum amplitudes of $i_1(\omega t)$ when varying
U$_\Psi$. They closely agree with the calculated
curves. This result shows that very low amplitude and phase errors are achievd over a
wide range of $\Delta\phi$.

CONCLUSIONS

The principle of single sideband homodyne
detection for independent measurements of
optical amplitudes and phases is described.
This technique can be extended by a double
phase modulation and by using a single sideband detector. This method utilizes integrated optic devices, and allows for an optical
network analysis by simple homodyne circuits.
When applied to fiber-optic sensors it has
the advantages that phase measurements are
not influenced by amplitude variations, and
that the stability mainly depends on the phase
stability of the modulation which can be
locked to a quartz oscillator.

REFERENCES

/1/ R.J. King, Microwave homodyne systems,
    Peter Peregrinus Ltd., 1978
/2/ J.S. Jaffee, R.C. Mackey, Nicrowave
    frequency translator, IEEE Trans. MTT-13,
    371 (1965)
/3/ O. Ostwald, B. Schiek, Network analysis
    by phase modulated homodyne detection,
    to appear in IEEE Transactions on In-
    strumentation and Measurement

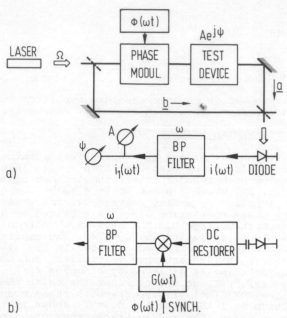

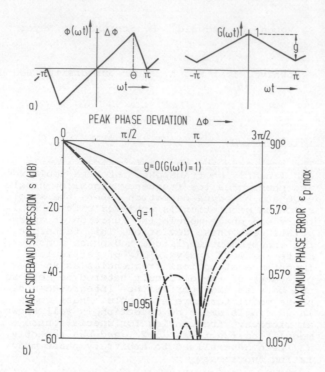

Figure 1 Principle of phase modulated single sideband detection (a), and introduction of an additional amplitude modulation (b).

Figure 2 Image sideband suppression and maximum phase error for a sawtooth phase modulation with θ=0.95π and amplitude modulation G(ωt).

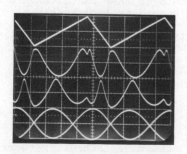

Figure 3 Integrated interferometer on LiNbO$_3$ (a), and signals U_ϕ, i(ωt), i$_1$(ωt) for θ=0.7π and U_ϕ = 12V$_{pp}$ (b).

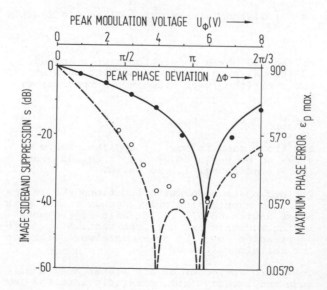

Figure 4 Theoretical and experimental suppression of the image sideband for θ=0.9π and g=0 (——•), g=1 (— — o).

OPTICAL BISTABILITY IN LiNbO$_3$ USING A MONOLITHICALLY INTEGRATED DIRECTIONAL COUPLER AND DETECTOR

R. Guglielmi, A. Carenco

Centre National d'Etudes des Télécommunications, 196 rue de Paris 92220 Bagneux, France

ABSTRACT

A pyroelectric waveguide detector has been realized on one output channel of an electro-optic directional coupler. The monolithic device, associated to an amplifier, has demonstrated optical bistability at 1.06μm.

Among the variety of systems used to achieve optical bistability, the hybrid circuit based on a directional coupler, a detector and an electrical amplifier, has demonstrated the attractive features of high efficiency and speed, and four-port availability (1,2). For such an application, LiNbO$_3$ is a very suitable material, due to its large electro-optic figure of merit, and the fact that low-loss waveguides can be formed in it by Ti diffusion. Towards a LiNbO$_3$ - based bistable circuit, we propose and demonstrate a detection scheme suitable for integration with the directional coupler.

It has been pointed out, that to perform loss measurements in LiNbO$_3$:Ti guides, the pyroelectric effect was a useful and sensitive technique (3,4). For detection purposes, it is necessary to make absorbing one of the output channels of the directional coupler. One simple solution is to take advantage of the large absorption which results from the metal cladding of a guide. This effect is particularly enhanced for the fundamental TM mode. In the Z-cut LiNbO$_3$ directional coupler, this mode is also conveniently the most efficient. The metallic absorption in turn heats up the crystal, giving rise to pyroelectric response.

The integrated device is schematically shown in Fig. 1. The Ti-diffused waveguides of the directional coupler are isolated from the electrodes by a SiO$_2$ layer which serves to prevent absorption. In one output channel, a Ti layer directly overlays the guide on a 1mm length. This induces a 40cm^{-1} absorption on the 1.06μm TM mode. When light is modulated, a pyroelectric voltage appears along the C axis of the crystal. The sensitivity is roughly estimated at 5 x 10^{-8} A/W.

The light output of the directional coupler can be monitored with the integrated pyroelectric detector on one channel and with an external detector on the other one.

The directional coupler switches from one state to the other with less than 8V. In order to achieve bistability, the electrodes of the directional coupler are fed-back with the pyro-signal, externally amplified with a JFET transimpedance first stage circuit (feedback impedance of 47 MΩ). Figure 2 displays the experimental demonstration of typical bistable behavior : the transmitted light power exhibits various hysteresis loops, as a function of the incident light (less than 5mw) depending upon the amplifier dc offset voltage.

In conclusion, a simple scheme has allowed to associate a directional coupler and a detector on the same LiNbO$_3$ crystal, and to achieve bistability. In other guided-wave active devices fabricated from ferroelectric crystals, such as LiNbO$_3$ or LiTaO$_3$, pyroelectricity offers a convenient way to detector integration. A large efficiency enhancement can be expected from impurities-induced photovoltaic effect (5). Further investigations are under way.

REFERENCES

1. P.S. Cross, R.V. Schmidt, R.L. Thornton and P.W. Smith, 1978, IEEE J. Quantum Electron., 14 577-580.

2. A. Carenco and L. Menigaux, 1980, Appl. Phys. Lett, 37, 880-882.

3. K.H. Haegele and R. Ulrich, 1979, Optics Lett., 4, 60-62.

4. A.M. Glass, I.P. Kaminow, A.A. Ballman and D.H. Olson, 1980, Appl. Optics, 19, 276-281.

5. A.M. Glass, D. Von der Linde, D.M. Auston and T.J. Negran, 1975, J. Electronic Mat., 4, 915-942.

AN OPTICAL SERRODYNE FREQUENCY TRANSLATOR

K.K. Wong and S. Wright

University of Glasgow, U.K.

INTRODUCTION

There is considerable interest at present in the use of optical fibres as sensors and transducers (Giallorenzi and Bucaro (1)). In these devices a long length of fibre is used to convert a small velocity perturbation due to some external influence into a measurable phase change. A number of schemes have been proposed for measuring this phase, generally using some form of interferometer. Some schemes employ heterodyne techniques to translate the phase change to a convenient intermediate frequency; this offers a number of advantages, including greater sensitivity and the prospect of a digital output (Cahill and Udd (2)).

In order to integrate any heterodyne scheme into a planar optical circuit, an optical frequency translator that operates upon light confined in a stripe waveguide is required, but such a device has not yet been reported. Acousto-optic methods that do not use lateral confinement have been explored for RF spectrum analysis and attempts have been made to extend this technique to stripe waveguides (Tsai et al. (3)). An electro-optic method based upon a waveguide Mach-Zehnder inter-ferometer has recently been proposed (Culshaw and Wilson (4)); this has a theoretical conversion efficiency of only 40%, and other sidelobes are suppressed to -16 dB compared to the translated component. We propose here a simpler technique, using a phase modulator driven with a sawtooth waveform. In principle, this can be an ideal frequency translator with negligible conversion loss. We have demonstrated this device in a Ti: LiNbO₃ waveguide, and achieved promising results.

PRINCIPLE OF OPERATION

Consider a phase modulator with an input optical wave of unit amplitude and circular frequency, ω_c, and with a signal $v(t)$ applied to the modulating terminal. The output wave will be of the form
$$\psi = \sin [\omega_c t + kv(t)]$$
where k is the modulator constant. If $v(t) = \omega_1 t/k$ (Fig. la),
$$\psi = \sin(\omega_c + \omega_1)t$$
and the output has been shifted in frequency to $(\omega_c + \omega_1)$. Due to the periodic nature of the sine function, an equivalent effect can be produced with an ideal sawtooth waveform, $S(t)$, of amplitude $2\pi/k$ and period $2\pi/\omega_1$ (Fig. lb).

This approach has respectable antecedents. It is similar to a method demonstrated at micro-wave frequencies, in which the electron transit time in a microwave valve was modu-lated with a sawtooth function (Cummings (5)). We take the name "serrodyne" from that work. The use of phase rather than transit-time modulation introduces significant but small differences, and we have achieved very similar results to the earlier work, but at optical frequencies.

In practice, the modulating waveform, $v(t)$ will depart from the ideal form $S(t)$. We have analysed the effects of a finite fall time and a small departure from a linear ramp, and outline the results here. The effect of a finite fall time can be calculated by considering the $v(t)$ represented by an asymmetric sawtooth (Fig. lc). With this modulating function, the largest sideband has a power $20 \log_{10}(\delta/T)$ relative to the translated frequency, at a frequency $(\omega_c - T\omega_1/\delta)$, and the sidebands close to $(\omega_c + \omega_1)$ are much smaller.

One form of departure from linearity can be represented by
$$v(t) = S(t) + a \sin \omega_1 t.$$

The effect of this deviation is to angle modulate the translated carrier, and produce the familiar Bessel function sidebands, of value $J_n(a)$ at frequencies $(\omega_c + \omega_1 \pm n\omega_1)$. For small values of a, only the first order sidebands are significant, at a level of approximately $20 \log_{10}(a/2)$ relative to the translated carrier. If the value of a represents a departure of 1 dB from the ideal value in $S(t)$, the sidelobes at ω_c and $(\omega_c + 2\omega_1)$ will be suppressed below -24 dB.

EXPERIMENT

The device was fabricated on a Y-cut plate of LiNbO₃. A 5 μm wide stripe of Ti, 300 Å thick, was formed on the surface photolitho-graphically. A waveguide was produced by diffusing this stripe into the crystal at 1000°C for 8 hours, using a closed tube method to prevent outdiffusion (Esdaile (6)). The resultant waveguide supported a single TE mode, and had a depth of about 2 μm. The Y-face was then mounted flush against the polished surface of a fused quartz block, and the X-faces were lapped and polished. After demounting, Al electrodes were fabricated on the device (Fig. 2). By calculation (Kaminow et al. (7)), the device should have a capacitance of 3.3 pF, and give a phase modulation of 0.97C radians/volt, where C is a correction factor depending on the waveguide geometry (0<C<1).

The device was tested by operating it in a Mach-Zehnder interferometer, shown in Fig. 3. The output from a Spectra Physics 120 laser (633nm) was divided into two beams at BS1, and two X40 microscope objectives were used to couple one beam into and out of the waveguide, in which it propagated as the TE_0 mode. The other beam was passed through a water filled acousto-optic Bragg cell operating at 22.2 MHz. The first order diffracted beam, shifted in frequency by this amount, was then combined with the device output on BS2, and the two beams directed on to a reverse biased HP4207 pin photodiode. It was noted that some unguided light passed through the crystal and could not easily be spatially separated from the waveguide output. The heterodyned product of the two beams was then observed on a spectrum analyser (HP8553B). The purpose of the Bragg cell was to simplify the interpretation of the device output spectrum. It translated this from dc to 22.2 MHz, so that positive and negative frequency components resulting from the serrodyne action could be distinguished.

With no signal applied to the device, the spectrum analyser displayed a single peak at 22.2 MHz. A sawtooth waveform with a period of 1 μSec was applied to the device from a Tektronix FG504 function generator. This waveform had a fall time of 20 nS, giving a ratio of $T/\delta \sim 50$. The voltage was adjusted to give a maximum signal at 23.2 MHz, which occurred at 8.8 volts pp. For a phase change of 2 π, this represents a modulation of 0.71 rad/volt, and a value of C = 0.74, which indicates a reasonable overlap between optical and modulating fields.

The resulting output spectrum is shown in Fig. 4. The translation efficiency to 23.2 MHz was better than -2 dB, and all other frequencies were suppressed to at least -22 dB. The peaks at frequencies slightly below the serrodyne sidebands are part of the spectrum of the modulating signal, radiated directly from the generator. The sideband levels observed are greater than those due to the finite falltime, so the spectrum of the modulating signal was examined directly. This was found to conform to that expected of an asymmetric sawtooth with 20 nsec falltime, but only within the measurement precision of ± 1 dB. However, deviations within this limit could well be responsible for the sideband levels observed. Another possible cause of the observed levels might be amplitude modulation within the phase modulator, but observation of AM was obscured by thermal noise and direct radiation from G1.

Other sources of spurious signal were identified directly. At the Bragg frequency of 22.2 MHz, a small signal due to direct radiation from generator G2 was observed. There was also a signal at this frequency due to the Bragg shifted beam and the unguided light through the device heterodyning on the photodetector. This signal fluctuated considerably, and was seen to be sensitive to mechanical vibration. It should be reduced by better coupling techniques. In particular, the use of a butt-coupled optical fibre for input coupling has been shown to reduce the unguided component considerably (Andonovic and Holbrook (8)).

The device was found to work as an efficient frequency translator over the range from 100 kHz to 5 MHz, only limited by the performance of the function generator. Translation to ($\omega_c - \omega_1$) was observed by reversing the polarity of the sawtooth, and translations to ($\omega_c + 2\omega_1$), ($\omega_c + 3\omega_1$) and ($\omega_c + 4\omega_1$) were observed by applying 4π, 6π and 8π radians peak phase change. The conversion efficiency decreased and the sideband levels increased as the order of the translation was increased, since the effect of any departure from S(t) is also increased.

CONCLUSIONS

A wideband optical frequency translator in which the light is confined in a stripe waveguide has been demonstrated. Good per-formance has been obtained with a modulating waveform obtained from a commercially available function generator. Moreover, the device is simple and easy to fabricate and will cause little excess loss when incorporated into a planar optical circuit. Work is in hand to define the performance limitations, but we expect that the present results do not represent the limit.

The maximum operating frequency for translation is limited by the capacitance of the device. If sidebands suppressed below -20 dB relative to the carrier are required, then the falltime must be less than T/10. Consequently, in a 50 Ω system, the present device should translate by a frequency up to at least 250 MHz.

ACKNOWLEDGEMENTS

The authors would like to thank their colleagues Mr. B. Bjortorp, Mr. I. Andonovic and especially Mr. R. Steele for useful discussions and also Mr. G. Beattie of Tektronix (UK) Ltd for his practical assistance.

This work is supported by the SERC. K.K. Wong would like to thank the University of Glasgow and the CVCP for their financial support.

REFERENCES

1. Giallorenzi, T.G., and Bucaro, J.A., April 27-29, 1981, "Fiber optical sensor technology", to be presented at the Third International Conference on Integrated Optica and Optical Fiber Communication, WI1.

2. Cahill, R.F., and Udd, E., 1979, "Phase nulling optical gyro", NAECON 1979, 8-13.

3. Tsai, C.S., Chang, C.L., Lee, C.C., and Liao, K.V., Jan. 28-30, 1980, "Acousto-optic bragg deflection in channel optical waveguides", Integrated and Guided-wave Optics, PD7.

4. Culshaw, B., and Wilson, M.G.F., 5th Feb. 1981, "Integrated optic frequency shifter modulator", Elect. Lett., 17(3) 135-136.

5. Cumming, R.C., Feb. 1957, "The serrodyne frequency translator", Proc. IRE, 45, 175-186.

6. Esdaile, R.J., 1979, Ph.D. Thesis, University of Glasgow.

7. Kaminow, I.P., Stulz, L.W., and Turner, E.H., 15th Nov. 1975, "Efficient strip-waveguide modulator", Appl. Phys. Lett., 27(10), 555-557.

8. Andonovic, I., and Holbrook, M.B., Private Communication.

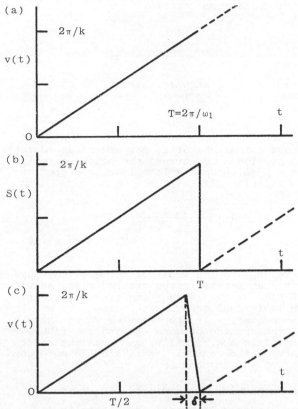

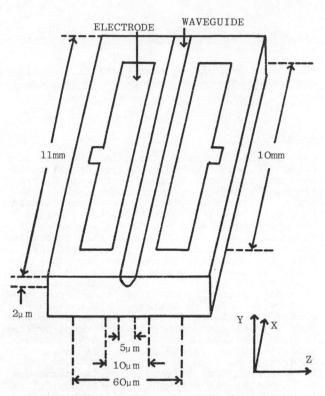

Figure 1 Modulating Waveforms (a) Ramp
(b) Ideal sawtooth (c) Sawtooth with
falltime δ.

Figure 2 The Ti:LiNbO$_3$ Phase Modulator

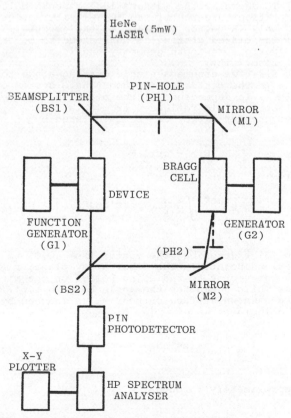

Figure 3 Experimental Arrangement

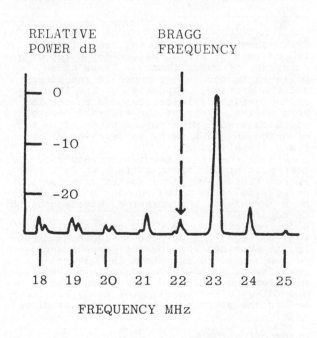

Figure 4 Experimental Results

PLANAR DEMULTIPLEXER WITH CHIRPED GRATING FOR OPTICAL FIBER SYSTEMS

T. von Lingelsheim and U. Unrau

Institut für Hochfrequenztechnik der Technischen Universität Braunschweig, Germany (FRG)

INTRODUCTION

Wavelength-division-multiplexing (WDM) is an effective means for multichannel operation by one optical fiber. Various 3-dimensional arrangements using lenses together with interference filters or gratings have been devised so far, e.g. (1)-(3). In 1979, the first planar approach to a multimode WDM was shown by Tangonan et al. (4). They suggested to couple input and output fibers to a film guide of which the opposite side was polished to a curve matching the phasefronts of the radiation from the input fiber. A plastic replica of a blazed grating was glued to the polish surface in order to introduce the necessary wavelength dispersion. Recently, Watanabe and Nosu (5) reported on a different construction uf such a device using a complicated 3-dimensionally curved blazed grating with three diferent groove angle sections.
The technological difficulty of Tangonans device consists in bringing the grating replica thoroughly in close contact with the polished surface. To avoid it, preliminary experiments with an available chirped grating glued to a flat polished surface have been undertaken (6). This seems to be a very promising idea, and it is the aim of this constribution to provide the theoretical means for construction of the whole device and for calculation of its performance.

CONSTRUCTION

Fig. 1 shows a scetch of the demultiplexer. It consists of a slab waveguide on top of a substrate of which front and rear sides are polished in order to accomodate the fibers and the grating, respectively. The thickness of the guiding film corresponds to the largest core-diameter of the fibers connected to it. Such a waveguide may be produced by ion-exchange in glass or by sandwiching different glass slabs (5).
In Fig. 1 the input fiber is attached to the left part of the front surface while three output fibers are glued to its right part side by side. (The actual number of output fibers depends on the number of wavelengths to be separated.)
A chirped grating is glued to the polished rear surface. Its purpose is twofold: firstly, it provides the correct phase transformation needed to transform the diverging phasefronts of the input beam in converging ones for the reflected rays which, secondly, according to the wavelengths used, are focused to the outgoing fibers by angular dispersion.

CHIRP CALCULATION

In Order to calculate the chirp function of the grating it is easier to think of the inverse problem where the interference pattern of twopoint sources is regarded in a plane whose normal vector is perpendicular to the connection line between the point sources. This problem occurs in holographic grating fabrication (7) and leads to

$$x_m = \frac{m\pi}{k} \left(1 + \frac{1}{\tan^2\alpha_o - (\frac{m\pi}{ky_p})^2}\right)^{1/2} \qquad (1)$$
$$m = 0, \pm 1, \pm 2, \ldots$$

Here, x_m are the locii of the grating steps along an x-axis in the grating plane as shown in Fig. 2. The index m denotes the number of an individual groove with respect to x=0 in the center, k is the material wavenumber, y_p is the spacing between grating and fiber array plane and, finally, α_o gives the angle between the central y-axis and the core center of the input fiber.

DEMULTIPLEXER CALCULATION

For the calculation of the intensity distribution in the fiber array plane several approximations are made:
a) Regarding the small dimensions of a single grating step in comparison to the total geometry locally homogenous plane waves are asumed to be incident to the grating;
b) To decrease the mathematical expense calculations are restricted to TE-polarization only (E-vector parallel to the grooves);
c) The blazed grating is understood to be of perfect sawtooth geometry and ideal reflectivity.
At first we assume a single grating stage to lie in the center of the coordinate system of Fig.2. In dependence of the blaze angle γ the electric field reflected from this minute rectangular mirror can be calculated by

$$\vec{E} = -\,\text{curl}\,\vec{F} \qquad (2)$$

from

$$\vec{F} = \frac{e^{-jkr}}{4\pi r} \int_A \vec{M}(\vec{r}')\, e^{jkr'\cos\xi}\,dA \qquad (3)$$

In (3) $\vec{r}$ describes the vector from (0,0) to various points in the fiber array plane, $\vec{r}'$ is the vector to radiating points on the mirror while ξ denotes the angle between $\vec{r}, \vec{r}'$. The unknown surface current $\vec{M}_s$ is induced by the incident wave

$$E_z^i = E_o e^{jk(x\sin\alpha_o + y\cos\alpha_o)} \qquad (4)$$

approximately through

$$\vec{M}_s = \vec{E} \times \vec{n} \quad \text{and} \quad \vec{M}(\vec{r}') = 2\,\vec{M}_s \qquad (5)$$

with $\vec{n}$ being the normal vector to the mirror plane. Now the solution of (2) and (3) is elementary. After transformation to sperical coordinates (r,ϑ,φ) the ϑ-component of the reflected field at $\vartheta = \pi/2$ is given by

$$E_\vartheta^r = j \frac{kabE_0e^{-jkr}}{2\pi r} \cos\gamma \sin(\varphi + \gamma) \, \text{sinc}(\varkappa) \qquad (6)$$

with

$$\text{sinc}(i) = \frac{\sin i}{i}$$

$$\varkappa = \frac{ka}{2} \left[\sin(\alpha_0 - \gamma) + \cos(\varphi + \gamma) \right]$$

where a and b are the dimensions of the mirror rectangle. The E_r - component vanishes and E_φ is negligible.
Next (6) has to be generalized for all small mirrors the actual grating consists of. Special attention has to be paid to the different phase relations for incident and reflected waves. Note that the angle α_m of an incident wave varies also with x_m (Fig. 2).
The final solution is computed from the superposition of reflections from all mirrors in the grating. The grating illumination is not uniform and depends on the far-field of the input fiber or another source attached to the input spot. For sake of simplicity a $\cos^2\varphi$ - distribution was assumed. The actual shape is not very important and restricts together with the N.A. mainly the size of the grating. The sum of reflected fields in the fiber array plane is given by

$$E_\vartheta^s = j\frac{kab}{2\pi} \sum_{m=-\infty}^{+\infty} \frac{AE_m}{R_m} e^{-jk\left[R_m - y_p\left(\sqrt{1+\tan^2\alpha_m} - \sqrt{1+\tan^2\alpha_0}\right)\right]} \qquad (7)$$

with

$$A = \cos\gamma_m \sin(\varphi'+\gamma)\,\text{sinc}\left\{\frac{ka}{2}\left[\sin(\alpha_m-\gamma)+\cos(\varphi'+\gamma)\right]\right\}$$

$$E_m = \cos\left[\pi\alpha_m/(2\text{arc}\sin(N.A./n_w))\right] E_0$$

$$R_m = \sqrt{(y_p/\sin\varphi)^2 + x_m^2 - 2y_p x_m \cot\varphi}$$

$$\varphi' = \pi/2 - \text{arc}\tan(\cot\varphi - x_m/y_p)$$

$$\alpha_m = \text{arc}\tan(\tan\alpha_0 - x_m/y_p)$$

For computation wavelength separation between 1.23μm, 1.27μm and 1.30μm was assumed. All fibers have 50μm/125μm core/cladding-diameters and N.A. = 0.2. The distance between fiber and grating plane is $y_p = 50$mm, the refractive index of the waveguide $n_w = 1.53$. The angle α_0 (Fig.2) is $\alpha_0 = 4^\circ$. (All this leads to an output fiber spacing of 150μm)
The chirped grating was dimensioned with a wavenumber of k= 7.569μm^{-1} corresponding to $\lambda_0 = 1.27$μm. Preliminarily, an optimal blaze angle is assumed at each point of the grating to show the theoretical limits of the device.

RESULTS

Thorough evaluation of (1) shows that the mean grid constant depends mainly on α_0. A variation of α_0 between 1° and 6° for a given y_p changes the mean grid constant between 27μm and 4μm, respectively. On the other hand the variation of y_p between 20mm and 50mm with α_0 held constant shows negligible differences for the central 200 grooves and becomes significant at the outer parts only. Due to the restricted illumination the grid constants at the last relevant grooves remain the same for all y_p. For the given data the grating constant increases from 6.73μm to 7.30μm over 1660 grooves from the center both ways. Near the center the chirp between adjacent grooves is very low, typically below 1nm. The dependance of the angular dispersion on α_0 is 0.09°/nm and does not depend on y_p. Fig. 3 shows the spectral demultiplexer curves. Crosstalk < -60dB, a 3dB-bandwidth of 0.7nm and dispersion losses < 0.2dB are found in this theoretical case. If the input fiber size is taken into account and radiation from incoherent point sources across the core is added up the output fiber illumination curves shown in Fig. 4 are obtained. In order to couple 90% of the incident energy to output fibers, their core diameter has to be enlarged by 15% with respect to the input fiber.

CONCLUSION

The given theory allows the calculation of multimode planar multiplexers and demultiplexers using gratings of arbitrary grating constant and blaze angle. Computations for a chirped grating show excellent properties for ideal chirp and blaze angle. Work taking deviations occuring in practice into account is in progress.

REFERENCES

1. Tomlinson, W.J., 1977,'Wavelength multiplexing in multimode optical fibers', Appl. Opt., 16, 2180-2194.

2. Aoyama, K., and Minowa, J., 1979, 'Optical demultiplexer for a wavelength division multiplexing system', Appl. Opt., 18, 1253-1258.

3. Watanabe, R., and Nosu, K., 1980, 'Optical demultiplexer using concave grating in 0.7-0.9μm wavelength region', Electron Lett., 16, 106-108.

4. Tangonan, G.L., et al., 1979, 'Planar multimode devices for fiber optics', Proc 5th Europ. Conf. on Opt. Comm., Amsterdam 21.5-1 to 21.5-7.

5. Watanabe, R., and Nosu, K., 1980 'Slab waveguide demultiplexer for multimode transmission in the 1.0-1.4μm wavelength region', Appl. Opt., 19, 3588-3590.

6. Tangonan, G.L., 1980, Private communication.

7. Unrau, U., and Nietz, R., 1980, 'Quick precision alignment of interferometric equipment', J. of Physics E: Sci. Instr., 13, 608-610.

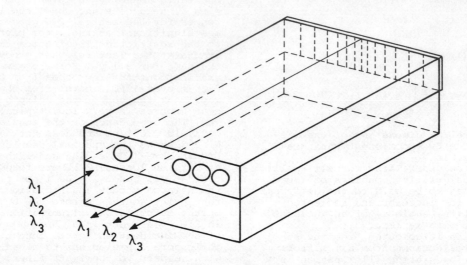

Figure 1 Principal configuration of the demultiplexer

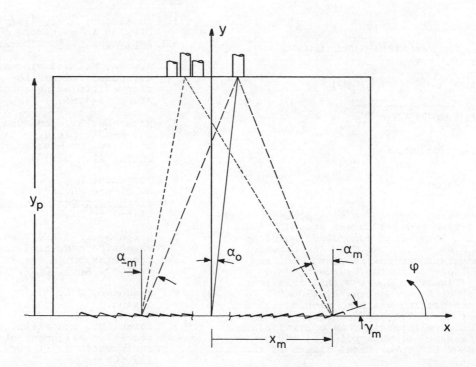

Figure 2 Geometrical definitions

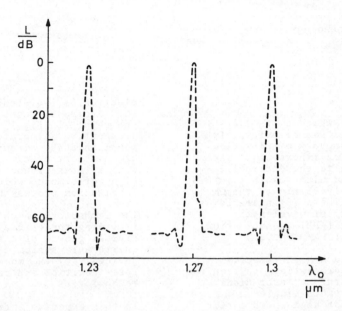

Figure 3 Spectral characteristic of the demultiplexer

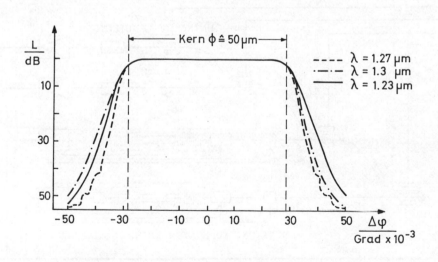

Figure 4 Output fiber illumination behaviour
(Attenuation L vs deviation $\Delta\varphi$ from the azimuthal angle φ at the output fiber axes)

INTEGRATED GaInAsP/InP LASERS

K. Iga and Y. Suematsu

Tokyo Institute of Technology, Japan

INTRODUCTION

The GaInAsP/InP laser which emits 1.2 - 1.7 μm of wavelength band has been intently studied since the discovery of ultra low loss behavior of silica fibers (1). The fact that we can utilize the entire 1.2 - 1.7 μm wavelength band with low loss (< 0.5 dB/km) if we completely eliminate the OH content from silica fibers (2) attracts us toward the possibility of wavelength division multiplexing (WDM) communication with ultra wide band ($\Delta\lambda/\lambda = 0.3$).

In order to realize such high capacity systems, very severe properties are required for laser diodes. First, a laser diode should operate in a stable single-longitudinal-mode under high-speed direct modulation which we call "dynamically single mode", especially for 1.5 - 1.6 μm wavelength. That is because the broadening of the lasing spectrum which is often experienced in a high-speed modulated conventional DH laser causes the pulse broadening due to the material dispersion of the optical fiber. Second, the lasing wavelength should be set to a desired one for the densely-packed WDM system.

Moreover, it is necessary to have an integrated optics which monolithically includes light sources, detectors, electronics, wavelength-tuning devices, modulators, switches, branches, and so on in one chip. In Fig.1 we illustrate a prospective integrated optics for a monolithic WDM circuit (3). The realization of this sort of circuit would provide us with i) mass-production of circuits and ii) high performance which comes from the fabrication process with no adjustment.

The technology which is indispensable for realizing these ideas must include (a) the crystal growth of the entire $Ga_x In_{1-x} As_y P_{1-y}$ quarternary system, (b) low threshold laser devices, (c) the monolithic formation of laser cavities and related fine technologies.

In this paper we introduce the effort in Tokyo Institute of Technology on integrated lasers associated with the GaInAsP/InP material. First, some structures of single mode lasers which have been demonstrated for the purpose of reducing the threshold current and stabilizing the mode behavior in the sense of "stationarily single mode" are reviewed, and second, we present the etching technique for fabricating waveguides and laser cavities and the recent development of the integrated twin-guide laser both for the purpose of obtaing single-longitudinal mode operation under high-speed modulation, i.e., dynamically stable single mode operation.

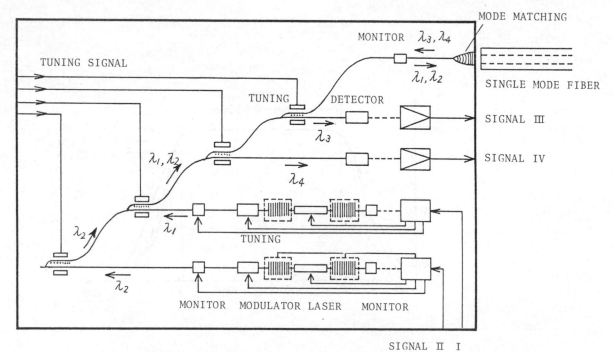

Figure 1   A prospective integrated optics for a wavelength division multiplexing system.

STATIONARILY SINGLE MODE LASERS

The transverse mode control which is employed with the help of index-guiding structure is very important to improve lasing properties such as modulation linearity and longitudinal mode behavior as confirmed experimentally and theoretically. Various types of GaInAsP/InP lasers with index-guiding which operate with single transverse and longitudinal mode have been realized under stationary operation. In this section, we present three types of mode-controlled GaInAsP/InP lasers which have been studied in the authors' group as illustrated in Fig.2.

Buried heterostructure (BH) lasers

The BH structure shown in Fig.2(a), which was first demonstrated in the GaAs system is prepared from a two step LPE growth process. We have fabricated a 1.5 - 1.6 μm wavelength BH laser (4), (5). First, the DH wafer is grown on a (100) InP substrate by a two phase solution technique. After depositing a SiO$_2$ film on the DH wafer by a CVD technique, narrow stripe masks are formed along the ⟨011⟩ crystal orientation. Narrow strip mesas are etched and then side bounding InP blocking layers are grown on the mesa-etched wafer resulting in the BH structure.

Threshold currents as low as 23 mA with the average of 42 mA are obtained from 84 samples with the cavity length of about 300 μm and the width of the active region of 3 - 5 μm. The lowest threshold current obtained is 28 mA and the highest differential quantum efficiency is 43 percent. Maximum output power of about 10 mW/facet is obtained under room temperature cw operation. For most samples, light output increases linearly up to twice the threshold. Transverse single mode operation of the BH laser is obtained up to more than three times the threshold.

Mesa-substrate buried (MSB) heterostructure lasers

The schematic structure of the MSB laser (6), (7) is shown in Fig.2(b). In the MSB laser, the GaInAsP active region is separately grown on the top of the mesa-shaped (100) InP substrate and it is completely embedded in the subsequent InP crystal by employing a single-step liquid phase epitaxy.

The MSB laser operates with a linear light output versus current characteristic over a range of more than three times the threshold. The output power is about 60 mW at 3.4 x I$_{th}$ for W = 10 μm, and 60 mW at 4.7 x I$_{th}$ for W = 5.5 μm, respectively. The minimum threshold current is 115 mA for a buried strip width of W = 5.5 μm. The external quantum efficiency of about 12 percent is observed at a current level of three times the threshold for both strip widths. A direct modulation characteristic of the MSB laser has been measured by a sharp current pulse method. The resonance-like peak which often appears around the cutoff frequency is not significant and the direct modulation up to about 2 GHz has been achieved. Recently, cw operation of the device with the current- confinement structure has been reported by Nomura et al with low threshold current of 20 mA for the emission wavelengths of 1.33 and 1.58 μm, (8).

Terraced-substrate (TS) lasers

The structure of a TS laser (9), (10) is shown in Fig.2(c). The waveguiding structure in the TS laser consists of a strip of non-uniform thickness active layer which provides us with effective index guiding and is prepared by single-step LPE growth. The active layer is not exposed to the air, therefore, as in the fabrication of the buried heterostructure. This is one of the merits of the TS laser in addition to the simplicity of process. The thickness of the active layer is 0.3 μm on the terrace slope and 0.2 μm on the flat regions, and the width of the strip waveguide is narrower than 3 μm.

Threshold currents of most TS lasers are 50 - 150 mA and the lowest value so far obtained is 44 mA in a laser with cavity length of 130 μm. A good linearity of the laser output power versus the injection current is attained above the threshold. The differential quantum efficiency of 42 percent for both facets and laser output power of 170 mW is obtained in the best laser. Cw operation has been also achieved at room temperature. Stable fundamental transverse mode operation is achieved by reducing the width of the strip waveguide to about 3 μm. The spot size at half maximum intensity is about 2 μm and does not change much with increasing current level, indicating that the transverse mode is stabilized successfully by the built-in index waveguide. Single wavelength operation is maintained up to 2.8 times the threshold in the best laser.

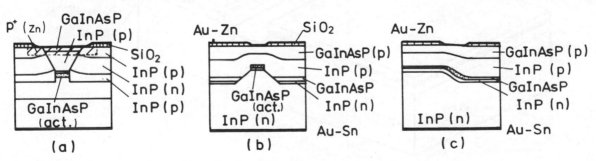

Figure 2   Stationarily single mode lasers with strip waveguides
            (a) BH laser   (b) MSB laser   (c) TS laser

DYNAMICALLY SINGLE MODE INTEGRATED LASERS

The monolithic formation of a laser
resonator which is free from the
conventional cleaving method is important
for integrating lasers into optical
circuits. This possibly provids us with
higher mode controllability than
conventional cleaved lasers. There have
been several methods such as etched mirrors,
distributed feedback or distributed Bragg
reflectors, integrated twin-guide (ITG)
structures, taper-coupled cavities and so
on. In this section, we introduce the
etching laser, and the ITG laser with
distributed Bragg reflectors for the purpose
of obtaining dynamically stable single
longitudinal mode operation.

Lasers with etched mirrors

The monolithic fabrication of etched-
mirror-facet lasers in the GaInAsP/InP
system provides us not only with the
capability of making very-short-cavity
lasers which could possibly lase in a single
longitudinal mode (11) and operate at low
current levels, but also of their monolithic
fabrication on the same substrate with other
optical or electronic components such as
monitoring detectors or driving field-effect
transistors and of easing the testing
process.

We have reported the fabrication of a
monolithically etched-facet GaInAsP laser
with oxide stripe geometry operating at 1.3
μm with one chemically etched (WCE) mirror
and one cleaved mirror (12). The groove is
etched in a batch process and one of the
walls of the groove is used as a laser
mirror. The other mirror is formed by
cleaving and it is possible to get two
etched-mirror lasers on one chip. This
arrangement also eases the problem of
mounting and handling of the laser chip.

Although the index-guiding structure is
required for stable single mode operation,
there has been a difficulty in etching a
vertical facet which is perpendicular to the
built-in index waveguide usually formed
along $\langle 011 \rangle$ direction in conventional BH or
TS lasers. We have succeeded (13) in making
index-guide structure, that is, a convex
lens-like active layer on a (100) InP
channeled substrate with grooves in $\langle 01\bar{1} \rangle$
direction by single step LPE growth as shown
in Fig.3. The active layer is 3 μm wide,
and the thickness of the center is 0.35 μm
followed by a uniform thickness layer of
0.25 μm. The etched facet by using KKI etch
(HCl:CH$_3$COOH:H$_2$O$_2$ = 1:2:1) is almost
vertical and smooth. We have fabricated
laser devices (λ = 1.3 μm) with one etched
mirror and one cleaved mirror with cavity
length of 100 μm - 300 μm. Threshold
currents of these lasers are around 160 mA
and single transverse mode operation has
been achieved. The light output vs. current
characteristic is linear. Expected
improvement of current confinement could
reduce the threshold current level of this
device. We have shown that the integrated
detector has sufficient sensitivity for
future use as a monitor to control the laser
(12). Mode controlled etched lasers would
provide us with an important technique to
fabricate short cavity lasers and in the
area of semiconductor integrated optics.

Distributed Bragg reflector (DBR)-ITG lasers

Distributed Bragg reflector (DBR) GaInAsP
injection lasers are thought to be promising
devices because they have the possibility of
wavelength selectivity, frequency stability
against temperature and injection current,
and wavelength control by the corrugation
period of the DBR. The stable single
wavelength operation of the laser even under
high-speed modulation is essentially
necessary in order to realize low loss and
wide band optical fiber communication
systems at the ultimate low loss wavelength
region, that is, 1.5 μm to 1.65 μm.

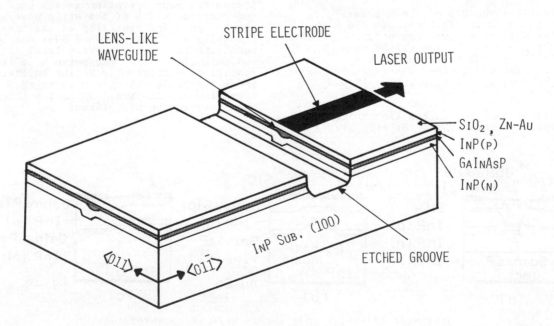

Figure 3   Etched laser with lens-like waveguide

Structure In DBR lasers, the corrugation which forms the DBR is fabricated at the different location from the active region, which gives the degree of freedom in the operation. Thus the following points are important: 1) efficient coupling between the active region and the DBR region and 2) the introduction of a low-loss waveguide on which the DBR is prepared. These two conditions can be satisfied in an integrated twin-guide (ITG) structure, which consists of the active guide and the output guide coupled to each other by the manner of a directional coupler. In this structure, the waveguide parameters are chosen so that phase velocities of the modes in both waveguides are equal. The DBR laser with the integrated twin-guide (ITG) structure for 1.5 - 1.6 μm of wavelength (14), (15), (16) is schematically shown in Fig.4. It consists of the twin-guide structure, that is, the active guide with an anti-meltback layer and the output guide coupled together. A thin anti-meltback layer is introduced to prevent the meltback of the GaInAsP active layer as described previously.

Fabrication The wafer is prepared by a conventional liquid-phase-epitaxy using the two-phase solution technique. First, seven quaternary and InP layers are successively grown on n-(100) InP substrate for 1.5 - 1.6 μm wavelength DBR-ITG lasers, respectively. The saturation temperature and the growth temperature of the active layer are 670°C and 615°C, respectively. The rather low cooling rate of 0.17°C/min is employed in order to obtain precise control of the growth thickness of each layer in the twin-guide structure. In the case of 1.5 - 1.6 μm of wavelength, an n-InP buffer layer (Te doped, 2.5×10^{18} cm^{-3}; $\cong$ 15 μm) is

grown first, followed by the successive growth of an n-GaInAsP output guide layer (the bandgap wavelength λ_{go}= 1.40 μm; Te doped, 5×10^{18} cm^{-3}; thickness b $\cong$ 0.55 μm), n-InP separation layer (Te doped, 5×10^{17} cm^{-3}; d $\cong$ 0.4 μm), undoped-GaInAsP active layer (λ_{ga} = 1.57 μm; a = 0.2 μm), p-GaInAsP anti-meltback layer (λ_{gam} = 1.35 μm; Zn doped, 1×10^{17} cm^{-3}; c= 0.1 μm), p-InP cladding layer (Zn doped, 1×10^{17} cm^{-3}; $\cong$ 3 μm), and p-GaInAsP cap layer (Zn doped, 3×10^{18} cm^{-3}; $\cong$ 0.3 μm). Those quaternary layers are lattice-matched to InP within 0.05 %.

Then, in order to leave narrow strip mesas the wafer is chemically etched down to the separation layer by using Br-CH$_3$OH and 4HCl:H$_2$O for GaInAsP and InP layers, respectively. In the second BH epitaxial growth, p- and n-InP blocking layers are grown in order to bury the twin-guide structure and define the current flow only through the stripe of 4 - 5 μm wide. In this process, the output guide layer located in the blocking zone is selectively melted back and buried in InP.

A part of the wafer thus prepared is chemically etched down to the separation layer to eliminate the active guide layer by using the selective etchants of 3H$_2$SO$_4$:H$_2$O: H$_2$O$_2$ and 4HCl:H$_2$O for GaInAsP and InP, respectively. First-order corrugations for distributed Bragg reflectors (DBR) are then formed on the etched surface of the exposed InP separation layer by the interference method by using 3250 Å line of a He-Cd laser and chemical etching technique. The typical period of the first order corrugation is 2363 Å for 1.525 μm wavelength.

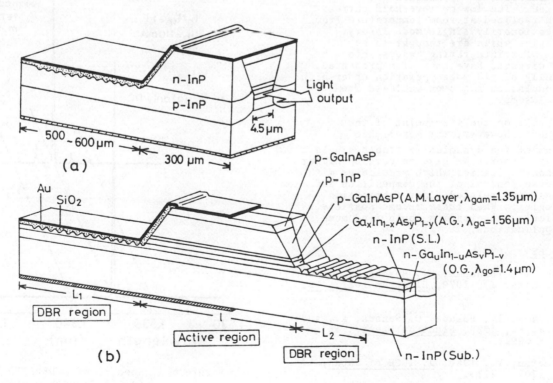

Figure 4   DBR-ITG laser with BH waveguide

74

Laser performance The GaInAsP/InP DBR-ITG laser has been tested under pulsed condition at room- temperature with a pulse-width of about 200 - 300 nsec. Stable single-longitudinal-mode operation has been maintained up to 1.39 times the threshold. Cw operation is also achieved at low temperature (16). The lasing wavelength varies with temperature by 1.25 Å/deg. for a specific device and mostly 0.8 - 1.25 Å/deg. The lasing wavelength is more stable against temperature compared with conventional type of GaInAsP DH lasers. The threshold current density is about 10 kA/cm² at room-temperature, which is due to the DBR wavelength deviation from the peak wavelength of the gain profile by about 500 Å. The differential quantum efficiency and the maximum output power are 11.7 % /output-facet and 25 mW, respectively.

Dynamically single mode operation High-speed direct modulation of the DBR-ITG device with a BH structure has been done with high frequency sine-wave current. Lasing spectra of a BH-DBR-ITG laser and a conventional BH laser modulated by 1 GHz sine-wave with 100 % modulation depth under a DC bias are shown in Fig.5 and Fig.6. In conventional DH lasers, it is known that rapid change of the injected carrier under the high-speed modulation causes spectral broadening as shown in Fig.5. For the DBR-ITG laser, however, the single-longitudinal-mode oscillation is maintained with a fixed wavelength due to the wavelength selectivity of the DBR, as shown in Fig.6. It has been also reported for the modulation by a short current pulse with 1.5 nsec pulse width (17).

SUMMARY

Three types of mode-controlled lasers with narrow strip waveguides are given and related fabrication techniques are summarized. The low cw threshold current has been realized at room temperature from these stationarily single mode lasers. Technologies which are thought to be necessary for integrating lasers into optical circuits have been also presented. Dynamically single mode operation under high speed modulation has been achieved from ITG-DBR lasers.

We are still on the startpoint of the development, however, for monolithic formation of the dynamically stable single wavelength laser. We have to develop higher performance epitaxies which provide us with the precise control of the composition and thickness, and related finer fabrication technologies. When these become possible, the optical communication could find much wider applications.

REFERENCES

1. Miya, T., Terunuma, Y., Hosaka, T., and Miyashita, T., 1979, Electron.Lett., 15, 106 - 108.

2. Moriyama, T., Fukuda, O.,Sanada, K., and Inada, K., 1980, Electron. Lett., 16, 698 - 699.

3. Suematsu, Y., 1980, J. IECE of Japan, 63, 1207 - 1213.

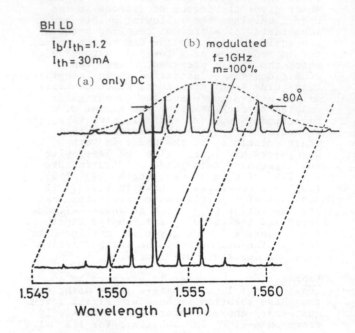

Figure 5   Spectra of a conventional BH laser
(a) Stationary operation
(b) Modulated

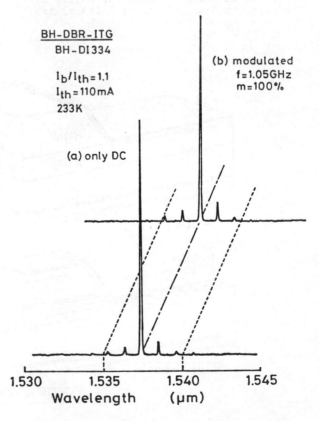

Figure 6   Spectra of a DBR-ITG laser
(a) Stationary operation
(b) Modulated

4. Arai, S., Asada, M., Suematsu, Y.,
    Itaya, Y., Tunbun-ek, T., and Kishino,
    K., 1980, Electron. Lett., 16,
    349 - 350.

5. Arai, S., Asada, M., Suematsu, Y.,
    Itaya, Y., Tunbun-ek, T., and Kishino,
    K., 1981, IEEE J. Quant. Electron.,
    Special Issue, to be published.

6. Kishino, K., Suematsu, Y., and Itaya,
    Y., 1979, Electron. Lett., 15,
    134 - 136.

7. Kishino, K., Suematsu, Y., Takahashi,
    Y., Tanbun-ek, T., and Itaya, Y.,
    1980, IEEE J. Quant. Electron., QE-16,
    160 - 164.

8. Nomura, H., Sugimoto, M., and Suzuki,
    A., 1981, Paper of Tech. Group, IECE of
    Japan, OQE-117.

9. Moriki, K., Wakao, K., Kitamura, M.,
    Iga, K., and Suematsu, Y., 1980, Jpn. J.
    Appl. Phys., 19, 2191 - 2196.

10. Wakao, K., Moriki, K., Kitamura, M.,
    Iga, K., and Suematsu, Y., 1981, IEEE J.
    Quant. Electron., Special Issue, QE-17,
    to be published.

11. Iga, K. and Takahashi, Y., 1978, Trans.
    IECE of Japan, E61, 685 - 689.

12. Iga, K., and Miller, B. I., 1980,
    Electron. Lett., 16, 830 - 832.
    Iga, K., and Miller, B. I., 1980,
    Electron. Lett., 16, 342 - 343.

13. Moriki, K., Iga, K., Uchida, M., Wakao,
    K., and Kunikane, T., submitted to
    Electron. Lett..

14. Utaka, K., Kobayashi, K., Kishino, K.,
    and Suematsu, Y., 1980, Electron. Lett.,
    16, 455 - 456.

15. Utaka, K., Kobayashi, K., and Suematsu,
    Y., 1981, IEEE J. Quant. Electron.,
    Special Issue, QE-17, to be published.

16. Kobayashi, K., Utaka, K., Abe, Y.,
    and Suematsu, Y., Electron. Lett. to be
    published.

17. Utaka, K., Kobayashi, K., and Suematsu,
    Y., 1981, IOOC'81, TuD5.

CURVED RIB WAVEGUIDES IN GaAs/GaAlAs

M.W. Austin

British Telecom Research Laboratories

INTRODUCTION

Much research effort is presently being devoted to single-mode optical fibre communication systems and on devices that may be integrated into optical circuits where the optical signals may be processed. Many of the devices likely to be incorporated in these circuits, such as directional couplers, Y-couplers and switches, may require sections of curved optical waveguide. Curved waveguides will also be necessary in order to increase the device packing density in integrated optical circuits.

The problem of continuous radiation from curved dielectric waveguides has been investigated by several workers (1-11), but very few curved guides have been fabricated (12-16). The guides that have been made were in either sputtered glass films on glass substrates or in lithium niobate and guiding was observed at a wavelength of $0.6328\mu m$. The aim of the present work is to investigate curved rib waveguides fabricated in III-V semiconductors, specifically the GaAs/GaAlAs and InP/InGaAsP material systems. This paper presents experimentally measured loss values for GaAs/GaAlAs curved rib waveguides.

CURVED WAVEGUIDE ANALYSIS

Curved dielectric waveguides have been analyzed by several workers using different methods of analysis (1-11). In most cases a perturbation analysis was used to determine the attenuation coefficient, α, of the curved guide due to radiation losses. The expressions given are valid only if the perturbation terms are small and this requires $\alpha R \ll 1$, where R is the radius of curvature of the guide. The analyses are also simplified by assuming that the refractive index difference between the waveguide core and the surrounding media is small.

In general, the attenuation coefficient is of the form:

$$\alpha = C_1 . \exp(-C_2 R) \qquad \dots\dots (1)$$

where C_1 and C_2 are independent of R. They depend on the difference in refractive index between the waveguide core and the surrounding media and hence the lateral confinement of the fields. The attenuation is most sensitive to the exponential whilst the more slowly varying multiplier reflects some of the geometrical details of the guide. Although the attenuation is generally found to be of the form given by Equation (1), Chang and Kuester (6,7), Kuester (8) and Lewin (9) predict a $R^{-\frac{1}{2}}.\exp(-C_3 R)$ dependence of the loss coefficient for all finite cross-section guide structures.

For GaAs rib waveguides with a small lateral effective refractive index step, ie guides with small rib heights, the optical

confinement is relatively weak and calculations show that a radius of curvature of the order of 2 mm is needed for the guide to exhibit a tolerable, less than 1 dB/rad, bending loss. To minimize the bending loss and reduce the radius of curvature it is necessary to fabricate guides which have a large rib height so that the rib is surrounded predominantly by air. This provides strong lateral confinement. For rib waveguides of this type, such as the structures shown in Figure 1, there is no analytical theory available for evaluating the bending loss. It is difficult to estimate the lateral penetration depth of the modal fields. A computer programme for plotting the modal field distribution, as calculated by a vector variational programme (17), is currently being developed and this will enable an estimate of the field lateral penetration depth and hence the expected radiation loss. For the type of rib structures being considered an extension of existing theories which are valid for weak confinement suggests that a very small radius of curvature is possible. However, for this case a large fraction of the guided energy may impinge on the etched side walls of the guide and the limiting loss of this type of guide may not be radiation loss due to bending but rather scattering loss due to rib wall imperfections.

EXPERIMENTAL

Curved rib waveguides to date have been fabricated on $GaAs/Ga_{0.82}Al_{0.18}As$ slices grown by MO-CVD. The GaAs layer is $0.8\mu m$ thick. A dry etch technique, ion beam milling, is being used to fabricate the waveguides. This is because most wet chemical etches are anisotropic and the resulting waveguide geometry depends on the guide orientation with respect to the semiconductor crystal axis. This is disadvantageous when 90° curved guides are being fabricated. A Veeco Microetch System is being used for the ion beam etching. Etching is done using 1 keV Ar^{+} ions at a current density of 1.2 mA/cm^2 for which the etch rate of GaAs is approximately 1600 Å/min.

The mask used to define the guides was made using electron-beam lithography and consists of 90° curves of 75, 100, 125, 150, 200, 250, 300 and $400\mu m$ radius with a $300\mu m$ straight section at each end of the curves. The curves are $3\mu m$ wide.

Optical waveguiding is observed at a wavelength of $1.15\mu m$ using a TM polarized He-Ne laser. The infrared radiation is 'end-fire' coupled into the cleaved end of the waveguide via a X45 microscope objective. The output cleaved face is imaged by an infrared camera via a X20 microscope objective. Both the input and output faces of the sample may be imaged in white light in order to locate the waveguides and to check the quality of the cleaved edges. As well as observing the guiding on a TV monitor, measurements of the mode intensity profile can be made by selecting and displaying a single camera line on an oscilloscope. This can be plotted on an X-Y recorder via a boxcar integrator. Loss measurements are made by imaging the waveguide near field onto a Ge photodetector and measuring the transmitted optical power. Measurements are taken at maximum output and no attempt is made to measure the power in the individual modes.

Results for two waveguide structures are shown in Figure 1. These guides support two modes. As can be seen, the insertion loss, measured as the incident optical power divided by the output optical power and not corrected for reflection losses at the ends of the wave-guide or input coupling efficiency, appears to increase exponentially for small radii and exhibits a minimum loss for radii of 250 to 300μm. For radii of curvature larger than 300μm it is believed that the increase in insertion loss is due to scattering loss. Scattering loss is proportional to guide length. The carrier concentration of the GaAs layer is approximately 2.10^{16} cm^{-3} and hence free carrier absorption losses are assumed to be negligible.

At 1.15μm, the wavelength of light being used in this work, the refractive indices of GaAs and $Ga_{0.82}Al_{0.18}As$ are approximately 3.44 and 3.35 respectively (18). The reflection coefficient at each end of the guides is therefore approximately 0.3. For the wave-guide structure with t = 0.3μm the input coupling efficiency was measured to be approximately 23%, or -6.3 dB. This value was obtained by measuring the insertion loss of short, approximately 250μm, straight guides. The poor coupling efficiency is because the minimum focussed laser spot size is approximately 2.5-3μm in diameter whereas the GaAs layer is only 0.8μm thick. This restricts the amount of light which can be coupled into the guide. A similar coupling efficiency is expected for the guides with t = 0.4μm. When corrected for reflection losses (-3.1 dB) and input coupling efficiency, a minimum loss of approximately 6 dB has been achieved for a 90° curved guide with a radius of curvature of 300μm. Most of this loss is thought to be due to scattering from the rib wall imperfections since existing theories which are valid for weaker confinement suggest that at 400μm the radiation loss due to bending the guide should be negligible. The rib profiles were studied with a SEM and the edge roughness appears to be of the order of 100 nm.

As can be seen, insertion losses are smaller for the guides with larger relative rib heights. This is due, for small radii of curvature, to a difference in the bending loss resulting from tighter field confinement. For larger radii, where the losses should asymptote to a value dominated by scattering loss, the 1-1.5 dB difference between the two waveguide geometries may be due to a loss occurring at the junction between the curved guides and the straight sections at each end of the curves. More work with guides of different geometries and material composition is needed to determine whether radiation loss, scattering loss, junction loss or input coupling efficiency is limiting the guide performance.

CONCLUSION

This paper has given a brief review of curved waveguide analysis and has presented some experimentally measured loss values for GaAs/GaAlAs curved rib waveguides. Results look encouraging. When corrected for reflection losses and input coupling efficiency, a minimum loss of approximately 6 dB has been achieved for a 90° curved guide with a radius of curvature of 300μm. In cases where the input coupling can be optimized, for example when a semiconductor laser and curved wave-guide are integrated on the same substrate, the insertion loss of these guides may be reduced significantly. More work is needed with guides of different geometries and material composition to determine the limiting loss mechanism.

The optimum material composition for tolerable radiation loss and single-mode operation is still being investigated. For efficient coupling to single-mode fibres a large guide width, of the order of 5-7μm, is required and to be compatible with directional coupler structures currently being investigated at BTRL, GaAlAs with a low aluminium content, 2-5%, is desirable. It is believed that single-mode curved rib waveguides with low radiation losses and radii of 200-300μm may be achieved. The limiting loss may not be radiation loss due to bending but rather scattering loss due to waveguide wall imperfections. Finally, the experience gained from investigating GaAs/GaAlAs curved rib waveguides is to be carried on to InGaAsP/InP structures.

ACKNOWLEDGEMENTS

The author wishes to acknowledge the assist-ance of Dr C.D.W. Wilkinson in providing the vector variational programme for calculating waveguide propagation constants and field distributions and Dr L.M. Walpita and Dr C. Riddle of the Department of Electronic and Electrical Engineering at University College, London are thanked for assisting with the ion beam milling of the curved waveguides. Mr E.J. Thrush of STL is thanked for growing and supplying the MO-CVD GaAs/GaAlAs material and Dr M.E. Jones and Mr C. Dix of BTRL are thanked for providing the electron-beam photolithography mask. Acknowledgement is given to the Director of Research, British Telecom, for permission to publish this paper.

REFERENCES

1. Marcatili, E.A.J., 1969, Bell Syst. Tech. J., 48, 2103-2132.

2. Marcuse, D., 1971, Bell Syst. Tech. J., 50, 2551-2563.

3. Kawakami, S., Miyagi, M. and Nishida, S., 1975, Appl. Opt., 14, 2588-2597.

   - correction, 1976, ibid, 15, 1681.

4. Takuma, Y., Kawakami, S. and Nishida, S., 1977, Trans. of Inst. Electron. Commun. Eng. Japan 60-C, 111-119.

5. Heiblum, M. and Harris, J.H., 1975, IEEE J. Quantum Electron., QE-11, 75-83.

6. Chang, D.C. and Kuester, E.F., 1976, Radio Sci., 11, 449-457.

7.   Kuester, E.F. and Chang, D.C., 1975,
     IEEE J. Quantum Electron., QE-11, 903-907

     - correction, 1976, ibid, QE-12, 371

8.   Kuester, E.F., 1977, Radio Sci., 12,
     573-578.

9.   Lewin, L., 1974, IEEE Trans Microwave
     Theory Tech., MTT-22, 718-724.

     - correction, 1975, ibid, MTT-23, 779.

10. Miyagi, M. and Nishida, S., 1978, J. Opt.
     Soc. Am., 68, 316-319.

11. Taylor, H.F., 1974, Appl. Opt., 13, 642.

12. Goell, J.E. and Standley, R.D., 1969,
     Bell Syst. Tech. J., 48, 3445-3448.

13. Goell, J.E., 1973, Appl. Opt., 12,
     729-736.

14. Goell, J.E., 1973, Appl. Opt., 12,
     2797-2798.

15. Furuta, H., Noda, H. and Ihaya, A.,
     1974, Appl. Opt., 13, 322.

16. Hutcheson, L.D., White, I.A. and
     Burke, J.J., Digest of Topical Meeting
     on Integrated and Guided-Wave Optics,
     Incline Village, Jan 1980, paper WB2.

17. Walker, R.G., 1981, Ph.D. Thesis,
     University of Glasgow.

18. Boyd, J.T., 1972, IEEE J. Quantum
     Electron., QE-8, 788.

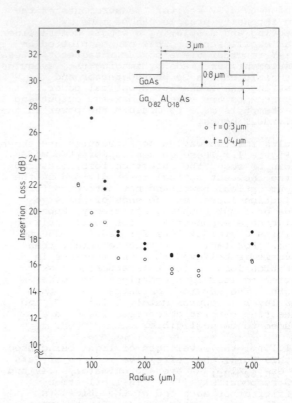

Figure 1.   Insertion loss as a function of
radius of curvature for the guide structures
shown in the inset.

PHOTOELASTIC OPTICAL DIRECTIONAL COUPLERS IN EPITAXIAL GaAs LAYERS AT 1.15µm

T.M. Benson, T. Murotani, P.N. Robson and P.A. Houston

University of Sheffield, U.K.

INTRODUCTION

Photoelastic waveguides which are useful
basic guiding structures for integrated
optical circuits have been fabricated in epitaxial layers of GaAs using evaporated metal
or SiO_2 stripes (1,2). Channel waveguiding
in an unclad epitaxial layer between two
parallel Schottky electrodes has also been
demonstrated in a GaAs electro-optic switch,
although the photoelastic effect was not
invoked as the confining mechanism (3).

This paper reports waveguide characteristics
in channel structures of the type shown in
Fig.1 for channel widths in the range 8 to
25µm. Optical waveguiding within the epitaxial plane results from the larger free
carrier plasma depression of the refractive
index in the GaAs substrate. Our calculations show lateral optical confinement is
caused largely by photoelastic effects.
Channels with widths between 12 and 20µm
show two well defined guiding regions and an
easily fabricated, highly synchronous, GaAs
electro-optical directional coupler switch
has been demonstrated using these strain
induced waveguides.

DEVICE FABRICATION

Gold films 0.5 - 2.0µm thick were evaporated
on to the GaAs layers and channels subsequently etched parallel to either the (011)
or (01̄1) crystallographic directions using
standard photolithographic and etching
techniques. The gold film is under tension
due to differential contraction between it
and the GaAs substrate on cooling after
evaporation. Strain fields are set up in the
GaAs and the resultant change in the relative
dielectric constant ε_r is the dominant cause
of waveguiding.

The material used in fabricating the devices
examined had a lightly doped GaAs epitaxial
layer ($|N_d-N_a| = 5 \times 10^{14} cm^{-3}$) 2.7µm thick,
grown on a heavily doped GaAs substrate
($N_d \sim 1 \times 10^{18} cm^{-3}$). After etching the wafer was
cleaved to expose the guide cross-section so
that light could be coupled in and out of the
device by 'end-fire' coupling.

Linearly polarised radiation from a 5mW,
1.15µm wavelength, HeNe gas laser was
focused on the cleaved input facet of the
waveguide. The near-field intensity
pattern was imaged on to an infra-red vidicon
and displayed on a TV monitor. The intensity
profile of a given TV line could be obtained
by sampling using a box-car integrator and
was plotted on an X-Y recorder.

EXPERIMENTAL RESULTS

Channels with widths of about 8µm in a 2µm
thick gold film showed one guiding region
directly under the centre of the channel for
both TE and TM modes in agreement with
calculated refractive index profiles. The
intensity attenuation coefficient was
measured as less than $1.5 cm^{-1}$ by sequential
cleaving. For channel widths of about 15µm
in the same thickness of gold film, two
guiding regions with their centres separated
by ~ 12µm were observed. By subsequently
cleaving and re-examining this device, the
output light intensity was determined as a
function of device length for various input
positions and with TE polarisation. These
measurements showed the device to be acting
as a directional coupler. Using well-known
equations for synchronous coupled guides (4),
a coupling length of (2.2 ± 0.2)mm was found
to be consistent with the experimental
results. The measured coupling length is
much shorter than reported by Campbell et al
(3) for couplers produced by having two
parallel narrow slots in close proximity.

By applying a bias to one electrode it is
possible to introduce a propagation constant
difference $\Delta\beta$ on one guide, through the
electro-optic effect (1,3) and hence alter
the amount of light in each guide at the output facet. Fig.2 illustrates the transfer of
light on application of a bias to one
electrode for a 1.9mm long device with a 2µm
thick gold film and channel width of 15µm.
By etching a 20µm gap midway along each
electrode and perpendicular to the guiding
channel, alternating $\Delta\beta$ couplers have been
successfully constructed.

TE and TM modes could be guided along the
edge of each electrode remote from the slot
in accordance with photoelastic theory. The
application of an electrical bias to the
Schottky electrode altered the propagation
constant of the guided TE modes only. In
order to measure the magnitude of the change
in propagation constant both TE and TM polar-
isations were equally excited at a remote
edge of one of the channel electrodes and
their output fields caused to interfere. By
application of a bias to this electrode the
electro-optic interaction with the TE mode
was measured. Fig.3 shows a plot of
$|\beta_{TE}-\beta_{TM}|L$ against applied bias for a 4mm long
sample. The graph follows closely a varia-
tion given by :

$$|\beta_{TE}-\beta_{TM}|L = \pi n^3 \frac{r_{41}}{\lambda_0} \frac{(V_B-3.4)}{d}L \quad \ldots\ldots (1)$$

$$= \Delta\beta L$$

where n is the layer refractive index, r_{41}
the electro-optic coefficient, d the epitaxial layer thickness, λ_0 the free-space wavelength, and V_B the applied bias voltage.

Using (1) to estimate the phase change between
the two coupled TE waves in the coupler on
the application of a bias to one electrode,
and a coupling length L_c of 2.4mm, a graph of
power (Fig.4) at the output of the input
guide for a 1.9mm long sample of width 14µm
was plotted as a function of bias using the

equation :

$$P_{input} = \cos^2 \alpha L + \frac{\Delta \beta^2}{4\alpha^2} \sin^2 \alpha L$$

where $\alpha^2 = \{C^2 + (\frac{\Delta\beta}{2})^2\}$, L is the sample length and $C = \pi/2L_C$ is the coupling coefficient. It can be seen that this plot and the experimental results presented in Fig.2 are consistent.

In channels of width larger than about 20μm the guiding became more complex with modes of higher order than the lowest symmetrical and assymmetrical ones required for directional coupler action.

NUMERICAL ANALYSIS

A numerical method has been developed based on a finite element technique (5) to enable the lowest order guided wave profiles and propagation constants to be calculated; this can conveniently be implemented on an HP85 desk calculator. The waveguide cross-section is divided into a rectangular mesh. The wave equation for TE modes takes the form (6) :

$$\frac{\partial^2 E_y}{\partial x^2} + \frac{\partial^2 E_y}{\partial x^2} + E_y^2 k_0^2 (n^2 - n_{eff}^2) = 0$$

where n_{eff} is the effective refractive index defined as β/k_0 where β is the propagation constant in the z-direction and $k_0 = 2\pi/\lambda_0$.

This is written in a five point finite difference form and the resulting set of difference equations are solved iteratively. Boundary conditions are forced to give a solution of the required symmetry and the conditions at infinity are approximated by containing the waveguide within an electric box. An initial guess at the fields is made and for each mesh point considered in turn a new field value is calculated. This new field value then overwrites the previous one. The eigen-value is improved after each complete scan by substitution into the finite element equivalent of the Rayleigh quotient (5). The procedure is continued until two consecutive values for the eigen-value agree to within a specified limit. Fig.5a,b show the normalised calculated profiles for the lowest symmetrical and anti-symmetrical TE modes 1.5μm beneath a 14μm wide slot with a force per unit length of 1200Nm⁻¹ at the metal discontinuities, corresponding to a 2μm thick gold film. The calculated photoelastic contribution to the dielectric constant at this depth is shown in Fig.6. The theoretical coupling length of 1.9mm obtained from the difference between the computed symmetrical and anti-symmetrical mode propagation constants agrees well with the experimental value.

CONCLUSION

We have reported novel directional coupler structures which have the advantage of strong coupling, easy fabrication and high synchronism without rigid fabrication tolerances. Experimentally measured coupling lengths of (2.2 ± 0.2)mm agree well with values calculated numerically.

REFERENCES

1. Westbrook, L.D., Robson, P.N., and Majerfeld, A., 1979, Electron. Lett., 15, 99-100.

2. Westbrook, L.D., Fiddyment, P.J., and Robson, P.N., 1980, Electron. Lett., 16, 169-170.

3. Campbell, J.C., Blum, F.A., Shaw, D.W., and Lawley, K.L., 1975, Appl. Phys. Lett., 27, 202-205.

4. Kogelnik, H., and Schmidt, R.V., IEEE J. Quantum Electron., QE-12, 396-401.

5. Davies, J.B., and Muilwyk, C.A., 1966, Proc. IEE, 113, 277-284.

6. Kogelnik, H., 1975, "Theory of Dielectric Waveguides", Tamir, T. (Ed.) :"Integrated Optics", (Springer-Verlag), 13-81.

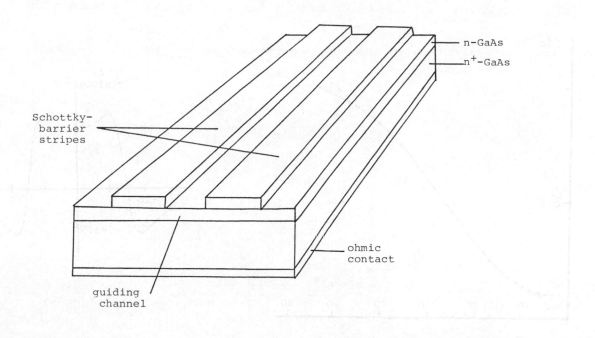

Figure 1 Schematic diagram of the photoelastic optical channel waveguide

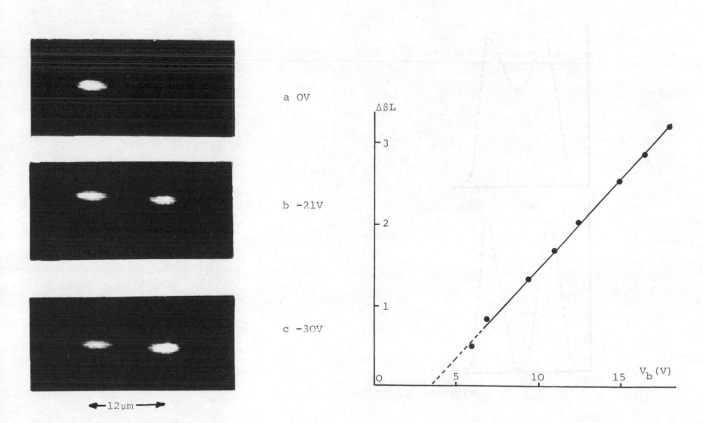

Figure 2 Near-field intensity pattern at the output of a 15μm channel. Bias on right-hand electrode

Figure 3 Plot of $\Delta\beta L$ against bias for an edge guided mode in a 4mm long sample

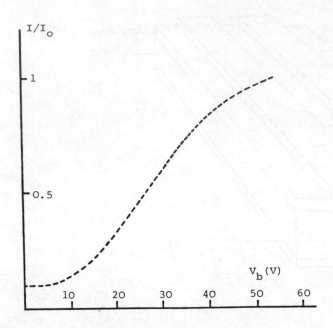

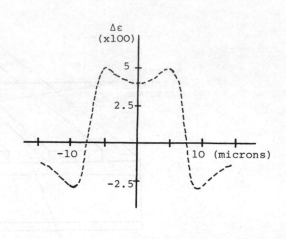

Figure 4 Calculated normalised power at output of excited guide against bias L_C=2.4mm, L=1.9mm, channel width = 15μm

Figure 6 Photoelastic contribution to dielectric constant at a depth of 1.5μm

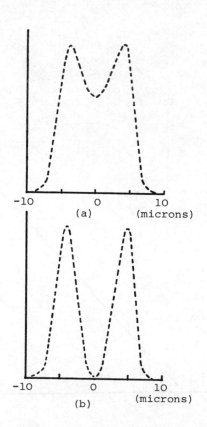

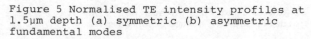

Figure 5 Normalised TE intensity profiles at 1.5μm depth (a) symmetric (b) asymmetric fundamental modes

MONOLITHIC INTEGRATION OF ACTIVE AND PASSIVE COMPONENTS IN GaAlAs/GaAs MULTILAYERS

N. Forbes, A.C. Carter and R.C. Goodfellow

Plessey Research (Caswell) Limited, Allen Clark Research Centre, Caswell, Towcester, Northants, U.K.

ABSTRACT

We have demonstrated the monolithic integration of edge-emitting LEDs and detectors coupled by a GaAlAs slab waveguide with GaAs FETs and resistors. The integration scheme utilises a GaAlAs multi-hetero-structure layer grown epitaxially onto a semi-insulating GaAs substrate which permits the separate optimisation of the materials structure for each type of integrated device.

INTRODUCTION

The capability of integrated optics becomes greatly enhanced if active processing electronics, light generators and detectors can be incorporated into monolithic integrated circuits along with waveguides and other passive elements.

Such components can be configured into circuits for many applications, including single chip fibre optic transmitters, receivers, repeaters, regenerators, linearised light sources, opto-isolators, PIN-FET detector front ends and electro-optical processers. All these applications would benefit from the reduced size, increased reliability and lower cost which can be realized with full component integration.

The functions which are essential to the listed applications are light generation, detection, guiding and electronics processing. In figure 1 is shown a scanning electron micrograph of a GaAlAs/GaAs chip which incorporates each of these functions, and so is a first step towards the optical/optoelectronic integrated circuit.

THE INTEGRATION SCHEME

In figure 2 we show schematically the layer structure which we have adopted and also illustrate how the various functions are formed. In this diagram Aluminium containing layers are shown cross hatched and GaAs layers are unmarked.

THE LED AND DETECTOR

The double heterostructure light emitting diode and photodiode are formed in layers (3), (4) and (5), the active layer (4) of each device being of the same composition (GaAs as shown). We found previously that discrete LEDs were efficient photodetectors for their own radiation when reversed biassed which justifies the use of identical device structures for light detection and emission. These experiments also showed optical coupling efficiencies around 10^{-5} for close proximity coupling of efficient LEDs and detectors across an air gap. To realise linearised light sources or optoisolators it would be advantageous to improve the coupling by some means to reduce the degree of electronic amplification needed.

WAVEGUIDING

To improve the optical coupling we have linked the LED and detector by means of an auxilliary slab waveguide. Half of such a scheme is shown in in figure 3(a). At normal operating currents radiation is emitted isotropically within the LED active layer. Various proportions are guided within the active layer or the lower guided layer depending on the compositions and hence refractive indicies of the layers. In figure 3(b) we plot the fraction of light guided in the 'auxiliary' lower guide layer as a function of Aluminium fraction in the layer. Using the auxiliary layer we have improved the optical coupling between integrated LED and photodetector from 10^{-5} to 10^{-3}.

An alternative scheme which has potentially higher coupling efficiency has also been considered. Light is guided from source to detector in the active layer which is continuous but electrically isolated by means of proton implantation. In practice the potential improvements may be realised only if the implanted regions are low loss.

THE FET

The high input impedance, high speed capability and the planar geometry of Gallium arsenide FETs make them attractive devices for integration. However correct FET operation is critically dependant on the doping profile and transport properties of the material, so dedicated FET layers were incorporated into the multilayer structure. The gate length used was 4μm which is close to the limit which would be achieved with the out of contact photoengraving necessary in our scheme because the FET is formed in the dopper layers 1 and 2. Optimisation of layer thicknesses will result in reductions in the minimum achievable gate length.

As shown in figure 4 two different FET geometries have been integrated and these had transconductances of up to 4mA/Volt for the 160μm gate width. This figure is about a factor of two lower than that expected for a FET of this design, and this can be attributed to non-uniform etching through the thick LED/Detector layers above the FET.

DISCUSSION

Integration schemes may be based on an insulating substrate [1] or a conducting substrate [2]. The former has the advantage of providing a conventional platform for a GaAs Schottky FET but several layers must be removed accurately to access the FET. The conducting substrate scheme involves the epitaxial growth of a semi-insulating FET buffer final layer which must be either removed or penetrated by localised diffusion to electrically access the under-lying layers.

We chose the insulating substrate scheme because it allowed the formation of waveguide coupled LEDs and detectors and mesa isolation which can give very low current leakage - an important feature for photo-detectors. The problems of controlled removal of material to access underlying devices can be reduced by the incorporation of Aluminium containing layers which can be readily selectively etched.

Present devices have been grown by a combination of two techniques. FET layers have been conveniently grown by chloride VPE, as this is well established and is capable of state of the art FET performance. LED/Detector layers have been grown using LPE, which results in efficient electroluminescent devices, but is not amenable to growth of more than ∿ 6 layers and is relatively inflexible when changes of layer structures are required.

Metal-organic chemical vapour deposition (MOCVD) has many advantages for multilayer monolithic devices. Large areas of uniform material can be grown, there is no limit to the number of successive layers which can be deposited and layers with high aluminium concentrations suitable for etch stops and waveguides can be incorporated. Multilayers have been grown by MOCVD which exhibit good FET and luminescence characteristics, unfortunately it has proved difficult to control lateral etching, which has prevented us fabricating complete optoelectronic chips by this technique. Improvements in compositional and doping uniformity will overcome these problems and open the way for considerably more complex structures than can at present be achieved by LPE.

The LEDs and detectors on the test chip design can be optimised for fibre coupling by the use of current confinement to reduce the emitting/detecting areas and by microlens coupling. The integration scheme is also amenable to the incorporation of laser devices using either cleaved or etched facets.

CONCLUSION

We have demonstrated the monolithic integration of most of the device functions necessary for the realisation of optoelectronic integrated circuits. The functions include light generation, light detection and amplification as well as resistive devices. The monolithic test chip incorporating these elements can be configured as a linearised source, PIN-FET detector or integrated repeater.

ACKNOWLEDGEMENTS

This work was sponsored by the Procurement Executive, Ministry of Defence. Thanks are due to R. Tyte of R.A.E. Farnborough for helpful discussions.

REFERENCES

1. M. Yust, N. Ber-Chaim, S.H. Izadpanah, S. Margalit, I. Ury, D. Wilt, A. Yariv. Appl. Phys. Lett. 35 (10) Nov. 1979.

2. T. Fukuzawa, N. Nakamura, M. Hirao, T. Kuroda, J. Umeda. Appl. Phys. Lett. 36 (3) Feb. 1980.

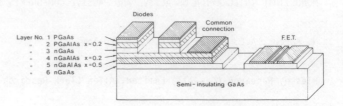
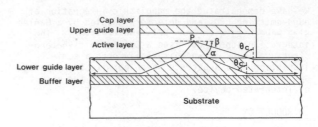

FIG.2 MATERIAL SPECIFICATION AND SCHEMATIC CROSS-SECTION

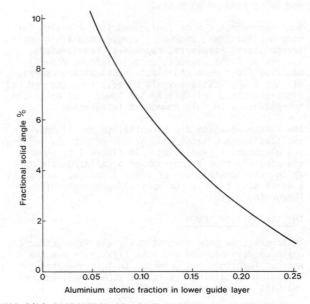

FIG.3(a) LIGHT TRAPPED IN LOWER GUIDE LAYER OF MESA'D
DOUBLE HETEROJUNCTION DIODE

FIG.3(b) PROPORTION OF LIGHT TRAPPED IN LOWER GUIDE
LAYER FOR BUFFER LAYER COMPOSITION x = 0.3

FIG.1 S.E.M. VIEW OF OPTOELECTRONIC TEST CHIP

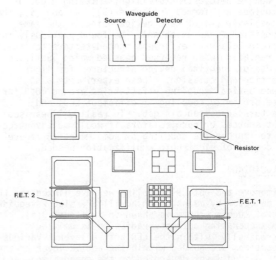

FIG.4 MASK LAYOUT FOR OPTOELECTRONIC CHIP

A NEW OPTICAL HETERODYNE DETECTOR WITH INTEGRATED DIFFRACTION GRATING

Yoichi Fujii, Hiroyuki Sakaki and Masayoshi Misawa

Institute of Industrial Science, University of Tokyo, Japan

A new optical heterodyne detector which has a grating coupler integrated on a photodetector is proposed, analyzed and experimentally demonstrated. The wavefront matching of the signal and the local waves is achieved by an integrated diffraction grating, which is directly fabricated on the surface of a Si photodiode. Because the coupling grating and the photodetector are integrated in a single body, this device has a simple structure and an excellent stability to the mechanical misalignment of the detector.

Suppose signal and the local plane waves are obliquely incident onto the grating as shown in Fig. 1. The wavefront matching condition of the heterodyne detection in this case can be derived from the momemtum conservation of the signal and the local plane waves on the plane waves on the plane of the grating. Thus the tangential component of the wave vector of the signal wave should be equal to that of the first-order diffracted local wave, as follows,

$$k(\sin\theta_s + \sin\theta_l) = 2\pi/\Lambda \qquad (1)$$

where θ_s and θ_l are the angle of incidence of the signal and the local waves, respectively, k is the wavenumber of both wave (note that their optical frequencies are approximately equal), and Λ is the period of the grating.

This equation determines the condition to obtain the maximum heterodyne output. The equation (1) holds for wavefront matching between the first-order diffracted signal wave and the refracted local wave. Consequently, the SNR can be ameliorated by the factor 2, since the outputs from both cases are added.

In the demonstrating experiment, the grating is fabricated on the SiO_2 coating of the Si photodiode, by the photolithography of the interference finges by the He-Cd laser ($\lambda = 442$ nm). The period of the grating Λ is 514 nm.

The schematic diagram of the heterodyne detection is shown in Fig. 2. The optical frequency is translated by the Doppler effect of the vibrating mirror on a loudspeaker. Changing the θ_s, the optimum θ_l is obtained as is in Fig. 3. This agree with the wavefront matching condition (Eq. 1)

The output power of the heterodyne detection is estimated by calculating a doefficient K of the coupling between the signal and the local waves by the grating. The K for the two wavefront matched wave pairs is given by

$$K = t(\theta_s)d(\theta_l) + t(\theta_l)d(\theta_s) \qquad (2)$$

where t is the transmission coefficient, and d is the first-order diffraction coefficient of the grating. After Tomlinson and Weber (1) the coefficient K for the shallow grating

($kh < 1$) is calculated. In Fig. 4, the relative value of the heterodyne output is compared as a function θ_l for the calculation and the experiment. Both values agree within the experimental error.

Next, we consider the effect of the wavefront mismatch due to the optical misalignment, which is a serious problem in the conventional optical heterodyne detection systems. When the difference of the incident angles between the signal and the local wave is $\Delta\theta$, the difference between both wave vectors is $\Delta k = k\Delta\theta$. The heterodyne output for the misaligned incidence is proportional to $|\text{sinc}(\Delta kL/2)|^2$, where L is the size of the photodetection surface.

In this heterodyne detector, allowance of the misalignment is same with conventional heterodyne detection. However, after the exact alignment of the signal and the local waves, the mechanical misalignment of this heterodyne detector is less effective. The wavefront mismatch Δk for the mechanical misalignment of this heterodyne detector can be calculated by using the vector momentum conservation law on the grating surface. The wavefront mismatch Δk for the small rotation $\Delta\theta_z$ around the z-axis is $(2\pi/\Lambda)\cdot\Delta\theta_z$. The Δk for the rotation around the y-axis is $k|\cos\theta_l - \cos\theta_s|\Delta\theta_y/\cos\theta_s$. The Δk for the rotation around the x-axis is $\sqrt{2}|\cos\theta_l - \cos\theta_s|\Delta\theta_x$. The latter two values of Δk are negligible by setting as $\theta_l = \theta_s$. The translations of this device have no effect. All these values of Δk should be corrected by the factor of $\cos\theta_s/\cos\theta_d$ in the photodetector medium of index n_1, as is shown in Fig. 1. In the experiment, the optical alignment of this heterodyne detector was much easier than the conventional heterodyne detector.

Since the incident signal and the local waves are usually Gaussian beams, the analysis mentioned here must be corrected. However, because the size of the detecting surface of the practial APD or the photodiode is about 200 μm, and because the confocal distance $\pi w_0^2/\lambda$ (where w_0 is the spotsize for $w_0 = 100$ μm is 30 mm ($\lambda = 1$ μm), both waves are practically plane waves if they are focused on the photodetector surface.

For smaller size (L $\sim$ Λ) of the photodetector, this type of the heterodyne detector becomes impractical, because the number of the grating lines on the photodetector is much smaller.

In conclusion, the optical heterodyne detector integrated with a grating, has shown to be a small-sized, alignment-easy heterodyne detector, and is applicable to the integrated detector in the communication and the measuring systems by the optical heterodyne detection.

The authors are very thankful to the helpful
discussions by Prof. Emerit. S. Saito and
Profs. J. Hamasaki and Y. Arakawa of the
Institute.

REFERENCE

1. Tomlinson, W.J., and Weber, H.P., 1973,
    J. Opt. Soc. Am, 63, 685-688.

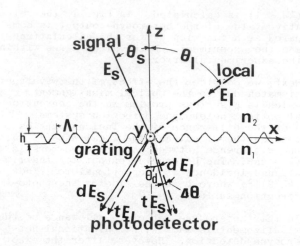

Fig. 1   The relation between the diffracted,
         and refracted signal and local waves

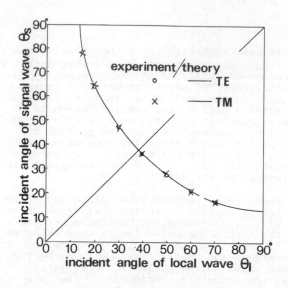

Fig. 3   Relation of the optimum heterodyne
         detection for incident angles of the
         signal and the local waves

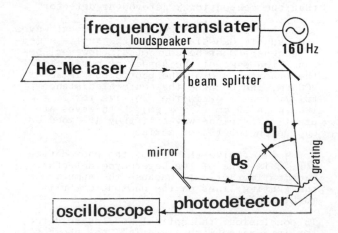

Fig. 2   Experimental set-up for the heterodyne
         detection by integrated photodetector

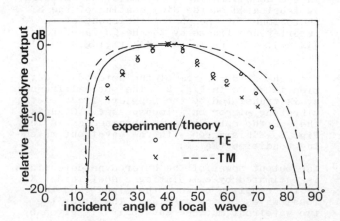

Fig. 4   Experimental and Theoretical relative
         heterodyne output as a function of
         the incident angle of the local wave

RECENT PROGRESS ON GUIDED-WAVE ACOUSTOOPTIC DEVICES AND APPLICATIONS

Chen S. Tsai

School of Engineering, University of California, Irvine, CA 92717, U.S.A.

INTRODUCTION

Guided-Wave Acoustooptics is one of the subareas of guided-wave optics that have received sustained interest in recent years (1). Presently there is even increased research and development activity on the resultant Bragg devices because of their immediate applications in real-time processing of wideband RF signals. Bulk-wave acoustooptic (AO) Bragg Cells had long been recognized as versatile interface devices between the RF signal and the optical wave in a conventional optical information processing system. Since the guided-wave counterpart was expected to play a similar role in future guided-wave optical information processing system, recent research and development efforts in guided-wave acoustooptics has been aimed at realization of efficient wideband planar Bragg cells in a $LiNbO_3$ substrate and feasibility demonstration of specific signal processing applications. In fact, this specific area of application has already provided an impetus and served as a focus for implementation of hybrid integrated optic signal processor modules. This paper reviews recent progress of this effort. Realization of efficient wideband planar Bragg cells together with some of the applications including spectral analysis and time-integrating correlation of RF signals, and high-scan rate high-resolution light beam scanning are discussed. A status report on AO Bragg deflection and switching in crossed channel waveguides, also in a $LiNbO_3$ substrate, is also given.

PLANAR ACOUSTOOPTIC BRAGG DEVICES AND APPLICATIONS

It is now possible to design and fabricate high-performance planar AO Bragg modulators and deflectors in Y-cut $LiNbO_3$ waveguides with a bandwidth approaching one GHz (1). For example, the deflectors of 700 and 470 MHz have been realized using multiple tilted-transducers (1,2) and a tilted-finger chirp transducer (2,3) respectively. A deflector which uses tilted phase-array transducers, aimed at 1GHz bandwidth, is also being pursued (4). Together with recent progress on fabrication of miniature laser sources, waveguide lenses, and photodetector arrays, hybrid integration of all of these components on a common $LiNbO_3$ substrate has become a reality. Clearly, the resulting integrated AO modules or circuits should find a number of unique applications in wideband multichannel optical communications and RF signals processing (1). One application that has already received a great deal of attention and interest is real-time spectral analysis of wideband RF signals (1,5-7). The next application expected to follow is time-integrating correlation of RF signals (8,9). In the latter application anisotropic Bragg diffraction was recently utilized to eliminate the need of imaging lenses, and thus greatly reduce the length of waveguide substrate (10).

CHANNEL ACOUSTOOPTIC BRAGG DEVICES AND APPLICATIONS

While planar AO Bragg devices have already reached

some degree of sophistication and found immediate applications, channel AO Bragg devices which may result from Bragg deflection in channel waveguides (Fig. 2) have only started to receive attention. The resultant channel devices are potentially more useful in fiber optic systems because of the compatibility in dimension and, thus, the relative simplicity in facilitating the coupling between the channel waveguide and the optical fiber. A recent experiment which employs a device configuration consisting of two crossed channel waveguides and one SAW transducer in a Y-cut $YiNbO_3$ has demonstrated simultaneously a high diffraction efficiency and a large deflector bandwidth (11). Such a device configuration should find a variety of unique applications in future integrated and fiber optic systems. In the application for heterodyne detection the frequency-shifted light can be conveniently used as a reference signal (local oscillator) in connection with optical communications and fiber optic sensing.

REFERENCES

1. Tsai, C.S., IEEE Trans. on Circuits and Systems., Vol. CAS-26, pp. 1072-1098 (Dec. 1979).

2. Lee, C.C., Liao, K.Y., Chang, C.L., Tsai, C.S., IEEE J. Quantum Electron., Vol. QE-15, 1166 (October 1979).

3. a. Liao, K.Y., Chang, C.L., Lee, C.C., and Tsai, C.S., 1979 IEEE Ultrasonics Symposium Proceedings, IEEE Cat. #79CH1482-9SU., pp. 24-27.

   b. Joseph, T.R., and Smith, W.R., Presented at Third International Conference on Integrated Optics and Optical Fiber Communication, San Francisco, CA., April 27-29, 1981, Technical Digest, pp. 122-123.

4. Stewart, C., Scrivener, G., and Stewart, W.J., Ibid, pp.122-124.

5. Hamilton, M.C., Wille, D.A., and Micceli, W.J., Proc. 1976 Ultrasonics Symp., IEEE Cat. #74CH1120-5SU, pp. 218-222.

6. Mergerian, D., and Malarkey, E.C., Presented at Third International Conference on Integrated Optics And Optical Fiber Communication, San Francisco, CA., April 27-29, 1981, Technical Digest, pp. 114-115.

7. Ranganath, T.R., Joseph, T.R., and Lee, J.Y., Ibid., pp. 114-115.

8. Yao, W.I, and Tsai, C.S., 1978 Ultrasonic Symposium Proceedings, IEEE Cat. #78CH1344-1SU, pp. 87-90.

9. Tsai, C.S., Wang, J.K., Liao, K.Y., SPIE, Real-Time Signal Processing II Vol. 180., pp. 160-163 (April 1979).

Invited Paper
*This work was supported by the AFOSR and NSF.

88

10. Lee, C.C., Liao, K.Y., Chang, C.L., and Tsai,
    C.S., Presented at 1980 IEEE Ultrasonics
    Symposium, Nov. 5-8, Boston, MA. Abstracts of
    Papers, p. 65.

11. Tsai, C.S., Chang, C.L., Lee, C.C., and Liao,
    K.Y., Post-Deadline Paper, 1980 Topical Meeting
    on Integrated- and Guided-Wave Optics, January
    28-30, Incline Village, Nevada, pp. PH7-1 to 4.

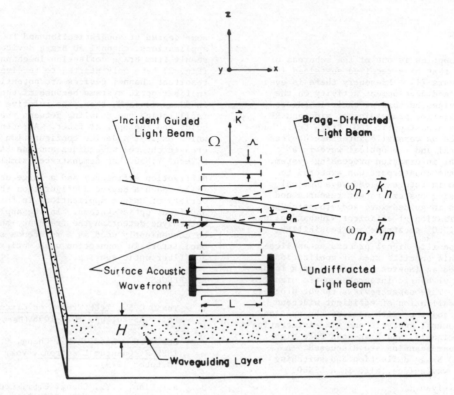

FIG. 1   GUIDED-WAVE ACOUSTOOPTIC BRAGG DIFFRACTION FROM A SINGLE SURFACE ACOUSTIC WAVE

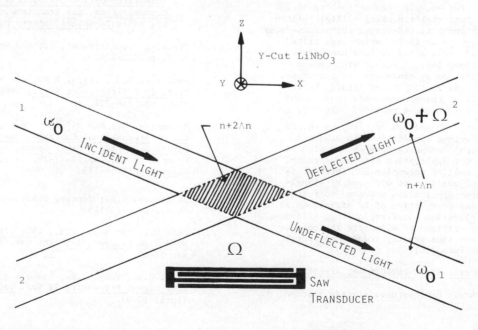

FIG. 2   ACOUSTOOPTIC BRAGG DEFLECTION IN CROSSED CHANNEL WAVEGUIDES

AN INTEGRATED ACOUSTO-OPTIC SPECTRUM ANALYSER USING GRATING COMPONENTS

V. Neuman, C.W. Pitt, L.M. Walpita

Department of Electronic & Electrical Engineering, University College London, U.K.

INTRODUCTION

The function of an acousto-optic spectrum analyser is to convert the frequency spectrum of an em wave pulse into spatially distributed light spots, the intensities of which are related to the amplitude weighting of the various frequencies in the pulse. The spectrum analyser [1], illustrated in Figure 1, has three basic components which perform as an expander of the narrow beam from the laser, converting it into a broad wavefront; an acousto-optic interaction region; and a Fourier transform lens. A current thin-film waveguide version of this device utilises geodesic lens components [2]; Luneburg lenses may also meet the requirements [3] of the device. However, geodesic lenses are difficult to fabricate, as are short focal length Luneburg lenses; in both cases non-planar processing steps are involved. Also, the lens index must be very much larger than the planar optical waveguide index, particularly for the Luneberg lenses.

In this paper, we present the use of grating components in the acousto-optic spectrum analyser; the components are fabricated by holographic and electron lithography processes. The conceptual layout of the device is illustrated in Figure 2.

The essential feature of the beam expander is that a narrow laser beam, passing through the grating strip, is slowly diffracted into a much wider beam. The expanded beam then interacts with a surface acoustic wave and is Fourier transformed by a Fresnel lens. We outline the techniques for fabricating the gratings and the current performance of the components.

BEAM EXPANDER

The condition for strong diffraction is that the incident beam impinges on the grating at the Bragg angle, Figure 3, given by,

$$\cos \theta = k_g/2\beta \qquad \ldots\ldots (1)$$

where β is the propagation constant of the incident wave, k_g the grating vector and θ is the angle between the incident field and the grating. The effect of a small perturbation will be to replace the original value of β by,

$$\beta_p = \beta + A\cos\left[k_g.x.\cos\theta\right] \qquad \ldots\ldots (2)$$

assuming a sinusoidal grating perturbation; A is a function of the grating depth, waveguide refractive index profile, and the type of mode. The diffraction efficiency has also been shown to depend on the term A which is a constant for a particular device. Thus, if A can be maximised, for example by increasing the grating depth, then the efficiency can be enhanced.

Fabrication technique

The grating may be fabricated in ion-exchanged glass [4], or diffused Lithium Niobate [5] waveguide. The waveguide forming process may be effected before or after the fabrication of the grating.

The holographic gratings were fabricated as follows. A high resolution resist was spun onto the substrate to a carefully controlled thickness, identified as the grating period. This condition minimises the effect of the standing wave formed in the thickness of the resist layer during exposure. After pre-bake, the photo-resist layer was exposed to interfering waves at a wavelength of 4579 Å, Figure 4. The laser beam was expanded and then converted into a plane wave; the wave was split into two plane waves by means of a cube beam splitter, and the two beams were then made to interfere at the appropriate angle. The exposed resist was developed, and baked, producing the grating profile in the resist layer. The grating was finally etched into the substrate by neutral ion-beam etching [6]. Since, in general, the light intensity entering a uniform depth grating beam expander decays as it diffracts into the expanded beam, the intensity of the expanded beam drops rapidly across the width of the grating. The amplitude uniformity may be improved by depth tapering the grating. To obtain the desired taper, we introduced a shutter in the neutral ion-beam etching machine; the shutter was mounted between the substrate and the collimated beam source. The movement of the shutter was controlled by a stepping motor under instruction from a microcomputer.

Performance

We have tested the grating in a glass waveguide by coupling light (6328 Å) into the guide with a prism coupler and adjusting the angle of beam incidence with respect to the grating so that the Bragg condition was satisfied. In order to measure the amplitude variation of the expanded beam, we probed the light scattered across the width of the beam, by means of a photodiode apertured to 600 microns. The pin-hole was approximately 1 mm away from the surface of the film. We compared the intensity profiles, Figure 5, of two expanded beams, one from a depth-tapered profile and the other from a grating without tapered depth. As expected, the grating of the uniform-depth beam expander exhibited a rapid drop in intensity across the beam width, curve (i). The linearly depth-tapered device shows better uniformity (curve (ii)) in output beam intensity. It will be noted that there is an intensity variation of sinusoidal form superimposed on the output beam in case (i). This is caused by depth modulation of the grating resulting from multiple reflections in the beam splitter used in the recording of the grating. At present, the best grating efficiency achieved is relatively low, currently 16% [7], but improvement is being sought by etching deeper gratings; alternatively, we might use guides with a larger film/substrate index difference.

GRATING LENS

Grating lenses are of two types: The Bragg type are thick lenses, while the Fresnel type are thin lenses. The Bragg lens may be designed to give 100% focussing efficiency, but the focussing efficiency drops rapidly when the beam illuminates the lens away from the Bragg angle. The Bragg lens, therefore, fails to satisfy the requirements of the spectrum analyser in which angular independence of the focussing efficiency is of prime importance. The Fresnel lens, on the other hand, exhibits a relatively low focussing efficiency; for example, a lens of 25% efficiency (half height spot size) has been demonstrated [8], although in principle the primary focus efficiency can reach 41% [9]. The attraction of the Fresnel lens is that it can be designed to retain a constant focussing efficiency over a wide angle of incidence. We have concentrated, therefore, on the Fresnel lens. The lens consists of half period zones so that a plane wave, diffracted by the lens, is focussed by constructive interference, see Figure 6. The converging light, emerging from the lens, should therefore have a cylindrical phase front. Referring to Figure 6, we may write,

$$\Delta\ell.\beta \;=\; z^2/2F.\beta \;=\; 2m\pi \quad \text{for } F>z \quad ....(3)$$

where β is the guided wave propagation constant, F the focal length, z the variable aperture and m an integer. This equation gives rise to a set of half period zones, (the even zones). The process of constructive interference occurs only if the remaining set of half period zones, (the odd zones), is such that the light is absorbed, or phase changed by π. The required Fresnel zone periodicity may be obtained by interfering a cylindrical and a plane wave; the recording system should be arranged so that the distance from the grating to the focal line of the cylindrical wave is equal to F, and the wavevector of the recording light is equal to β.

Fabrication techniques

Thin film Fresnel lenses have been fabricated at UCL both by electron beam lithography and by the type of holographic interference techniques indicated below. In order to fabricate the grating, the zone material, (usually a thin film metal, or dielectric, overlay), was sputtered or evaporated onto the substrate and then coated with the resist to a thickness less than the minimum zone periodicity. The holographic system is essentially a Michelson interferometer operating at 4579 Å wavelength and shown in Figure 7. The argon-ion laser light was expanded and converted to a plane wave. The plane wave was split into two beams; one beam was reflected from a plane mirror and the other from an aluminium coated cylindrical lens. The two beams were then recombined to form the interference pattern. The resist was exposed to the interfering plane and cylindrical wavefronts on the normal to the substrate plane. The developed and post-baked resist has a sinusoidal depth variation, illustrated in Figure 8a. In order to replicate the half period zone in the metal or dielectric film on the substrate, we ion etched the structure as indicated in Figure 8b; the etching process was completed chemically, Figure 8c. After removing the remaining photo-resist, the lens was reduced to the required width, Δx, (see Figure 6), by a second photolithography process. The resulting zone structure is shown in Figure 9a. The poor acuity of the zone edges is caused by vibration and particle defects in the recording system. We have also written EBL lenses at the Rutherford and Appleton Laboratory's EBL facility. In this case, the Fresnel zones were well defined, (Figure 9b). However, to date, it has proved difficult to produce large aperture (>2 mm), short focal length lenses, (focal length < 1 cm), by EBL since the zone periodicity is less than 2 micron at the edge of the lens. The present electron beam process has difficulty in writing lines which are so close because of the so-called proximity effect.

An absorption type lens may be made by ion-exchange or diffusion after the definition of the zone sections. In this case, there is no guiding layer formed under the zones, i.e. the guiding layer is discontinuous in these regions. The phase lens, on the other hand, has been made by depositing the zone pads after the waveguide has been formed. In this case, the lens pad thickness, (Δt - see Figure 6), is controlled so that a π phase shift condition is satisfied. Since the width, (Δx) of the lens is small, usually less than 50 microns, we may use metal pads as phase shifting Fresnel zones; the metal absorption is relatively small (< 0.1 dB per 50 microns).

Lens performance

We measured the spot size of the Fresnel lens focus by the technique suggested by Chang et al [8]. A cylindrical lens was employed to focus the laser (6328 Å) onto a rectangular prism base so that the light coupled into the waveguide was an expanded plane wave. To minimise phase distortion, we found it necessary to ensure very good contact of the prism edge to the film at the point where the light coupling was effected. We then coupled light out of the film, after the focus of the Fresnel lens, using a second rectangular prism. The focal point of the lens was then imaged using a x 40 microscope objective lens. We measured the size of the focal spot by scanning the Fresnel lens focus, at the image plane of the x 40 lens, with a photodiode apertured to 25 microns. The traces for an EBL absorption type lens and an absorption type interference lens are shown in Figure 10. In the case of the interference lens, the measured 3 dB spot size, (3.5 microns), is larger than that theoretically expected, (2.5 microns). Further, the measured signal/first side-lobe ratio (measured as circa 3dB) is very much smaller than that theoretically suggested, (15 dBs). In the case of the EBL lens the measured half-height spot size (2.0 microns) is very much closer to the theoretical value, (2.4 microns) and also the signal/first side-lobe ratio (measured as 9 dBs, theoretical 14 dBs), is much closer. The improved performance of the EBL lens is the result of the improved acuity of the pad edges.

CONCLUSIONS

We have demonstrated that gratings may be used effectively as components in the spectrum analyser. The beam expander is an attractive component since it is a relatively straightforward matter to improve the efficiency close to the theoretical maximum of 100%. The efficiency of the Fresnel lens may also be improved, at the cost of reducing the lens resolution, by including the first side-lobes. This would produce a component efficiency of approximately 32%. We are currently engaged in integration of the components, and hope to report on this procedure in the near future.

ACKNOWLEDGEMENTS

The authors wish to thank Professor E.A. Ash for the original concept of the grating beam expander and for subsequent discussions concerning this component.

The work has been supported by the Ministry of Defence (PE). We are grateful to Dr. Brian Wheeler and Dr. Paul Williams for their suggestions and comments during the programme.

REFERENCES

[1] Anderson, D.B., Boyd, J.T., Hamilton, M.C., and August, R.R., "An Integrated Approach to Fourier Transform", IEEE J. Quantum Electron., 1977, QE-13, pp. 268-275.

[2] Chen, B., Marmom, E., and Morrison, R.J., "Diffraction Limited Geodesic Lens for Integrated Optic Circuits", Appl. Phys. Lett., 1978, 33, pp. 511-513.

[3] Yao, S.K., and Anderson, D.B., "Shadow Sputtered Diffraction Limited Waveguide Luneburg Lenses", ibid., 1978, 33, pp. 307-309.

[4] Neuman, V., Parriaux, O., Walpita, L.M., "Double-alkali Effect: Influence on Index Profile of Ion Exchanged Waveguides", Electron. Lett., 1979, 15, pp. 704-706.

[5] Naitch, H., Nunoshita, M., Nakayama, T., "Mode Control of Ti Diffused $LiNbO_3$ Slab Optical Waveguide", Applied Optics, 1977, 16, pp. 2546-2549.

[6] Singh, S.P., and Pitt, C.W., "Neutralised Ion-beam Etching", Proc. Ion Plating & Allied Techniques Conference, July, 1979, pp. 37-46.

[7] Neuman, V., Pitt, C.W., Walpita, L.M., "Guided Wave Holographic Grating Beam Expander - Fabrication and Performance", Electron. Lett., 1981, 17, 4, pp. 165-167.

[8] Chang, W.S.C., and Ashley, P.R., "Fresnel Lenses in Optical Waveguides", IEEEE J. Quantum Electron., 1980, QE-16, pp. 744-754.

[9] Collier, R.J., Burckhardt, C.B., Lin, H.L., "Optical Holography", New York, Academic Press, 1971.

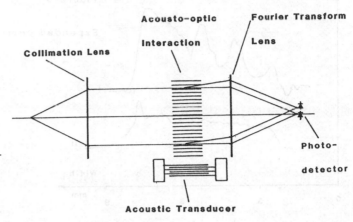

Figure 1   Spectrum analyser configuration

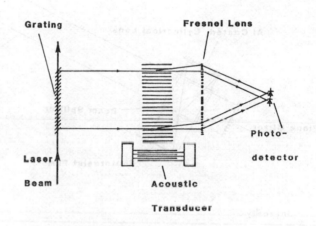

Figure 2   Grating version of the analyser

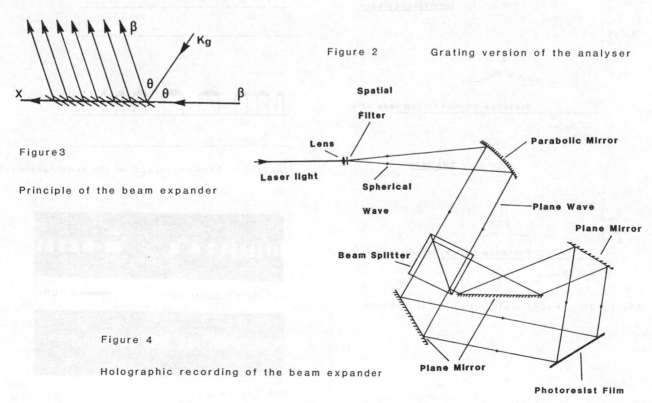

Figure 3

Principle of the beam expander

Figure 4

Holographic recording of the beam expander

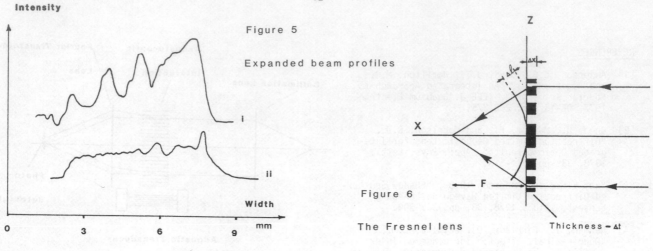

Figure 5

Expanded beam profiles

Figure 6

The Fresnel lens

Thickness = Δt

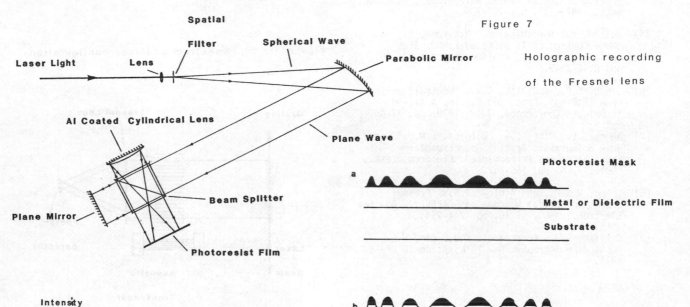

Laser Light

Lens

Spatial Filter

Spherical Wave

Parabolic Mirror

Plane Wave

Al Coated Cylindrical Lens

Beam Splitter

Plane Mirror

Photoresist Film

Figure 7

Holographic recording of the Fresnel lens

Photoresist Mask

Metal or Dielectric Film

Substrate

Figure 8   Process steps in the lens fabrication

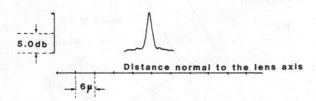

Interference Lens

Distance normal to the lens axis

3.5db

8μ

EBL lens

5.0db

Distance normal to the lens axis

6μ

Figure 10   Focal distribution of the lenses

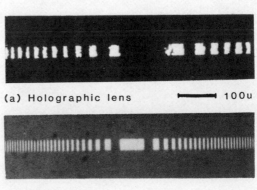

(a) Holographic lens   100u

(b) EBL lens

Figure 9

MICROWAVE FREQUENCY ACOUSTO-OPTIC GUIDED WAVE DEVICES

C Stewart, G Scrivener and W J Stewart

Plessey Research (Caswell) Limited, Allen Clark Research Centre,
Caswell, Towcester, Northamptonshire NN12 8EQ

INTRODUCTION

The Bragg interaction of guided optical waves with surface acoustic waves (SAW) has been the subject of extensive investigation in recent years (1). The application of such devices (known usually as acousto-optic Bragg cells) to r.f. spectrum analysis and correlation has provided the main impetus for research. Much effort has gone into the fabrication of 'integrated' models (2,3) in order to demonstrate the feasibility of the technology. This technology is based on titanium (Ti) diffused optical waveguides on lithium niobate (LiNbO$_3$) substrates. A solid state laser source and silicon detector array are both coupled to the waveguide with geodesic waveguide lenses to collimate and transform the guided light. These devices have been directed at providing a useful, if somewhat modest, specification. Typical parameters are 400 MHz bandwidth with a frequency resolution of 4 MHz (ie 100 channels) and a dynamic range of 30-35 dB. This performance may be contrasted with current bulk wave acousto-optic device performance of 1 GHz bandwidth with 1 MHz resolution and a noise limited dynamic range of about 55 dB (4). However, bulk wave devices require drive power levels usually in excess of 1 Watt while the integrated version requires only a few tens of milliwatts. The overall aim of the work reported here is to realise with guided optical wave-SAW devices the performance levels achievable using bulk wave devices. A particular design goal is to develop a guided wave Bragg cell with 1 GHz octave bandwidth and 1 MHz resolution.

Such performance levels demand improvements in all aspects of the 'integrated' system. SAW transducers capable of operating in the 1-2 GHz band while maintaining relatively low drive powers are required. Waveguide lenses must be capable of diffraction limited performance with high efficiency at guided beam apertures of about 4 mm. Wide dynamic range detector arrays with close element spacing which can be readily integrated to the optical waveguide are also required. In our work we have initially addressed the problems of achieving low waveguide scattering in Ti-diffused LiNbO$_3$ waveguides and the investigation of wide bandwidth acousto-optic Bragg cells fabricated in these waveguides at centre frequencies above 1 GHz and bandwidths up to 500 MHz.

TECHNOLOGY CONSIDERATIONS

Operation of SAW devices over octave bandwidths at frequencies in excess of about 1 GHz places severe demands on the substrate material in terms of acoustic attenuation, the lithographic process used to define the transducer pattern and the electrical matching network needed to achieve efficient transducer operation. It is fortunate that lithium niobate, already well established as an acousto-optic medium at UHF frequencies, exhibits a sufficiently low acoustic attenuation to allow useful operation of SAW devices at least up to 2 GHz. Nonetheless the effect of acoustic attenuation must be considered from the outset for acousto-optic devices in this frequency range. For YZ LiNbO$_3$ (Ycut, Zpropagating SAW) attenuation is about 1 dB/µs at 1 GHz. Since attenuation follows an approximately frequency squared dependence, at 2 GHz the value is about 4 dB/µs. For 1 MHz resolution in a 1 GHz bandwidth the transit time of the acoustic beam across the optical aperture is 1 µs. If the near edge of the optical beam is about 1 mm from the transducer ($\sim$0.25 µs) then the acoustic signal is reduced by 1.25 dB and 5 dB for 1 GHz and 2 GHz signals respectively at the further edge of the optical beam. Transducer losses also tend to increase with frequency further compounding the effect. Since it is desirable for many applications to have a reasonably flat acousto-optic response some form of input or output signal weighting must be considered.

In order to fabricate the submicron linewidth transducer patterns appropriate to the 1-2 GHz frequency band we have chosen a direct write electron beam lithography approach. Transducer patterns at frequencies up to 1.7 GHz have been produced. The electron beam approach was preferred since optical lithography is near its operating limits at the linewidths required (<0.5 µm). Also the flexibility of direct pattern writing allows different device geometries to be investigated with short fabrication delay. Two different procedures have been developed based on positive and negative electron resist techniques respectively.

Perhaps the most important requirement for high frequency guided wave devices concerns the substrate surface quality. A good optical quality finish is crucial for successful SAW transducer fabrication at submicron linewidths, for maintaining the lowest possible acoustic attenuation and for achieving low optical waveguide scattering which is necessary for good dynamic range of the acousto-optic device. Our starting material is optically polished lithium niobate substrates manufactured by Crystal Technology, which exhibits an excellent surface finish. After Ti-diffused waveguide fabrication no difference in the surface quality is apparent for the best waveguides.

DESIGN CONSIDERATIONS

External parameter specifications such as centre frequency, bandwidth and drive power are directly related to the characteristics of the acoustic wave and hence the transducer design while dynamic range and frequency resolution are mainly dependent on optical beam characteristics. The usual requirements are for wide (octave) bandwidth with low drive power, wide dynamic range and fine frequency resolution. An inherent trade-off exists between acousto-optic efficiency and bandwidth since high efficiency demands long interaction lengths (wide acoustic apertures) while wide bandwidth requires narrow apertures. Additionally, choosing a particular centre frequency for the transducer leaves the acoustic aperture as the only remaining adjustable parameter to achieve a particular value of input impedance, which is usually 50 Ω. We have adopted a transducer geometry which allows good acousto-optic efficiency without requiring narrow apertures for wide bandwidth. The basic transducer type is the frequency controlled beam steering array transducer (5) which in our experiments is designed for minimum 20% fractional bandwidth. Wideband acousto-optic performance is achieved by using multiple tilted transducers of this type, as described by Tsai (1). These first order beam steering transducers allow phase matching of the acoustic propagation constant to within 1% of the optimum Bragg angle condition over the full bandwidth.

The width of the optical beam determines the frequency resolution for a given bandwidth. Using a Rayleigh criterion for resolution implies that an optical aperture of nearly 4 mm is required for 1 MHz resolution within a 1 GHz bandwidth. Such beam widths require that the Ti-diffused waveguide is of good uniformity. Wide dynamic range is only possible if the optical waveguide exhibits little low angle scattering. Scattering from both the main beam to produce an unwanted background and between channels which introduces crosstalk must be avoided. At higher frequencies the problem of scattering from the main beam to the diffracted beam positions is reduced somewhat due to the increased Bragg angle, eg. at 1 GHz $2\theta_B \approx 4.5^{\circ}$. Our best waveguides exhibit scattering levels at least 50 dB down on the main beam level at 1° from the main beam direction. This scattering performance is equal to the sensitivity of the measurement system.

EXPERIMENTS

Ti-diffused waveguides were formed by r.f. sputtering titanium metal onto lithium niobate substrates and then placing in a tube furnace for 10 hours at 1050°C. Typical initial titanium thickness was 200 Å and oxygen flowed through the furnace tube at a rate of 1 litre per minute throughout the diffusion time. The substrates were supported at the top of an unsealed boat loaded with lithium niobate fragments. A single sample from each batch was assessed for waveguide scattering quality using the method described by Vahey (6). All samples were visually inspected before passing to the electron beam fabrication facility. Details of the fabrication process for high frequency SAW devices will be described elsewhere (7).

To date three series of transducer have been investigated. Initial experiments were performed at 1 GHz centre frequency with a single 200 MHz bandwidth device. Both mode converting ($TE_0 - TE_1$) and non-mode converting ($TE_0 - TE_0$) diffraction was observed. This device was also used in a demonstration experiment at the European Space Agency's laboratory in Noordwijk, Holland, with an input signal buried in noise. The device was successfully interfaced with a computerised integration system for the extraction of low-level signals. Thus, in this configuration the device was operated at a constant average power level with simultaneous excitation over the full device bandwidth.

The second series of device consists of two tilted beam steering transducers with a composite bandwidth of about 500 MHz centred at 1.35 GHz. Initial experiments with these devices have been described previously (8). The third series uses a single beam steering transducer of 500 MHz bandwidth centred at 1.7 GHz which has also been designed as part of a multi-element array. Results obtained with both types will be presented at the conference and compared with design parameters.

This work has been carried out with the support of the Procurement Executive, Ministry of Defence, sponsored by DCVD and the European Space Agency.

REFERENCES

1. Tsai, C S, 1979, IEEE Trans, CAS-26 (12), 1072 (and references therein).

2. Mergerian, R, et al, 1980, Appl Optics, 19, 3033.

3. Chen, B, et al, 1980, Paper ME3, OSA Meeting on Integrated and Guided Wave Optics, Jan 28-30.

4. Itek Technology product data.

5. De la Rue, R M, et al, 1973, Elect Letts, 9, (15) 326.

6. Vahey, D W, 1980, Paper TuD4, OSA Meeting on Integrated and Guided Wave Optics, Jan 28-30.

7. Brambley, D, et al, to be published.

8. Stewart, C, et al, Paper WK2, IOOC '81, April 27-29, San Francisco.

F.T. GEODESIC SYSTEM FOR HIGH-RESOLUTION SPECTRUM ANALYSER

S. Sottini, V. Russo, and G.C. Righini

Istituto di Ricerca sulle Onde Elettromagnetiche, C.N.R., Italy

INTRODUCTION

The implementation of an IO spectrum analyser for RF signals has already received a noticeable attention (1-3). According to Fig.1, let us consider a typical analyser employing an AlGaAs laser source, a $LiNbO_3$ substrate, a SAW Bragg-modulator driven by the RF signal, and two geodesic lenses. Such a device should be particularly well suited to achieve frequency resolutions in the range 0.5 to 5 MHz with operating bandwidth up to 1 GHz. The projects which are currently being developed, however, pursue much more limited goals: resolution 4 MHz, and bandwidth 500 MHz (2,3). An increase of the bandwidth can be achieved by using, in parallel, two analysers covering contiguous bands. On the contrary, the resolution capability can be improved only with much larger substrates and/or smaller detector element spacing, according with the following, well known equations (1):

$$D = g \, Q \, v/\Delta f \qquad (1)$$

$$S = \lambda \, \Delta f \, F/n_{eff} \, v \qquad (2)$$

The first one gives the lens diameter D as a function of the resolution Δf and of the SAW velocity v. This equation includes a factor Q describing the optical lens quality (equal to the actual spot size divided by the diffraction-limited spot size), and a truncation factor g of the beam. The other equation interrelates the spacing S of the detector array with the resolution Δf and the focal length F of the Fourier-Transform lens. λ is the free space optical wavelength, and n_{eff} the effective refractive index of the propagating mode.

Let us refer to Fig.1 and suppose $\lambda = 0.87$ μm, $n_{eff} = 2.173$, $v = 3500$ m/s, $S = 12$ μm, two geodesic lenses with $Q = 1.5$ and $g = 1.41$ (e^{-1} truncation). It can be easily shown that, in order to have a frequency resolution of 1 MHz, the focal length F must be at least 104.6 mm, and the total length of the substrate should be 140 mm or longer. The realization of such a substrate would be difficult and costly.

Here we suggest an alternative optical system constituted by two geodesic lenses, able to perform a magnified FT of the input signal. For a given spectral resolution, it gives rise to a drastic reduction of the length of the device.

THE OPTICAL SYSTEM

Let us suppose that the FT lens of a spectrum analyser is requested to have a focal length F in order to achieve the desired resolution. Referring to Fig.2, and chosen a magnification coefficient m, it can be seen that the same resolution capability can be obtained by using a focusing lens of focal length F/m together with a second lens, centered at distance p from the axial focus of the first one, which gives a magnified image of the signal spectrum at distance q= mp. The angular aperture of the second lens must be equal or greater than that of the first one. The two geodesic lenses can be easily designed following the general method already described by Sottini et al.(4).

The substrate size saving assured by the FT two-lens system has been evaluated for different values of m and S. Fig.3 shows the amount of size saving for four values of the spacing S: 7.5, 12, 14, and 20 μm respectively. In particular, with a magnification coefficient m = 5, the same specifications of the example previously considered ($\Delta f = 1$ MHz, bandwidth 500 MHz) could be obtained with a substrate size of 80x15 mm^2, which is acceptable.

As to the optical performance of such a system, two effects must be considered, which are related to the rotation symmetry of the geodesic lenses. The first effect is the curvature of the imaging line. In the example above considered, if we used a straight-line detector array, we would notice a maximum displacement between image points and detectors of the order of 1.25 mm. Consequently an optical beam having about 30 μm spot size would impinge on the detector array, lowering the resolution in an unacceptable way. Therefore a curved array detector should be used. The second effect is the spherical aberration due to the fact that the second lens must image point sources that are not at the same distance from its axis (Fig.2). Some computations have shown that this effect is absolutely negligible: in our example, the maximum amount of transverse spherical aberration is of the order of 1.5 μm.

With the aim of a qualitative investigation of the practical difficulties one meets in the realization of the two-lens geodesic system, we have planned to fabricate some prototypes in our workshop. The fabrication technique at our disposal is the same already used to realize other geodesic components (Righini et al.(5)); however, in the present case the fabrication errors of the two lenses add one to the other to give a total effect which is particularly dramatic.

A first realization is shown in Fig.4, having the following characteristics:
- focusing lens L_1: diameter 12 mm, central depth 1.675 mm, linear aperture 10 mm, focal length 21.7 mm.
- imaging lens L_2: diameter 8 mm, central depth 2.251 mm, linear aperture 6 mm, focal length 3.75 mm, object distance p = 4.5 mm, and image distance mp = 22.5 mm.

The BK-7 substrate was covered with an epoxy resin film (Araldit MY757) doped with Rh B to visualize the guided light. A laser beam, prism-coupled into the film and deflected by the two lenses, is clearly visible in Fig.4. However, as expected, this prototype has shown unacceptable focal shifts (of the order of 3-4 mm for each lens).

In order to reduce the consequence of the fabrication errors, we think of evaluating the focal length of the first lens immediately after its grinding and then positioning the second one. The focal length can be calculated with a good precision from the measurement of the lens central depth by means of the simple rule of thumb, reported by Sottini et al.(6).

CONCLUSIONS

The geodesic optical system here described seems to represent an attractive solution for the implementation of integrated optical spectrum analysers with relatively high resolution (1 MHz), without heavily increasing the substrate size or reducing the spacing S of the detector array. Due to the rotation symmetry of the geodesic lenses, the detector array usually must be positioned on a curved line.

An attempt is being done of fabricating a prototype of the FT two-lens geodesic system with the rough technique at our disposal in the workshop. We think that, in spite of the high fabrication errors, interesting results could be obtained by evaluating the actual focal length of the first lens immediately after its fabrication, and then repositioning the substrate for the grinding of the second lens.

REFERENCES

1. Hamilton, M.C., Wille, D.A., and Miceli, W.J., 1977, Opt. Engin., 16, 475-478

2. Barnoski, M.K., Chen, B.U., Joseph, T.R., Lee, J.Y.M., and Ramer, O.G., 1979, IEEE Trans. Circuits & Systems, CAS-26, 1113-1124

3. Mergerian, D., Malarkey, E.C., Pautienus, R.P., Bradley, J.C., Marx, G.E., Hutcheson, L.D., and Kellner, A.L., 1980, Appl. Opt., 19, 3033-3034

4. Sottini, S., Russo, V., and Righini, G.C., 1979, J.Opt.Soc.Am., 69, 1248-1254

5. Righini, G.C., Russo, V., and Sottini, S., 1980, Proc. S.P.I.E., vol.235

6. Sottini, S., Russo, V., and Righini, G.C., 1979, IEEE Trans. Circuits & Systems, CAS-26, 1036-1040

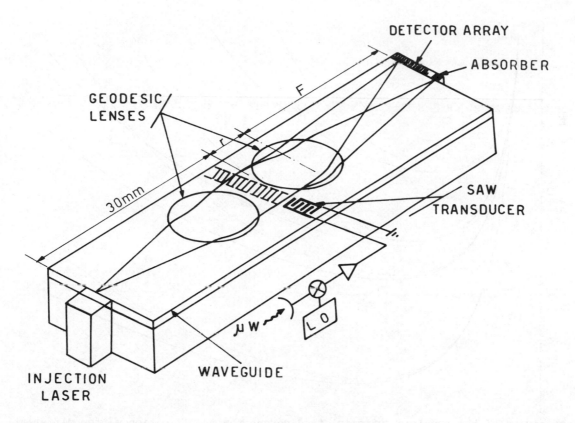

Figure 1 Sketch of a conventional IO spectrum analyser having two geodesic lenses on a LiNbO$_3$ substrate. F is the focal length, and r is the depression radius of the FT lens.

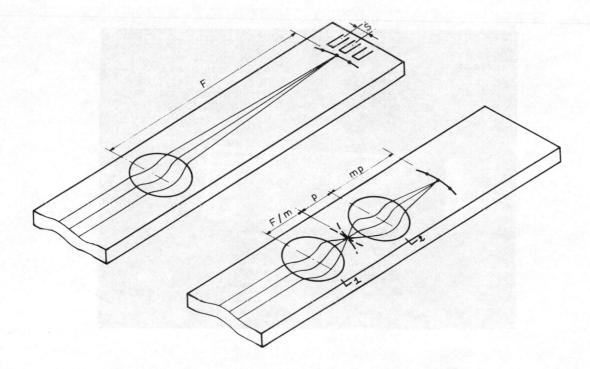

Figure 2 The usual FT lens of a spectrum analyser (top) is compared with a system which consists of two lenses (bottom) and assures a substrate-size saving.

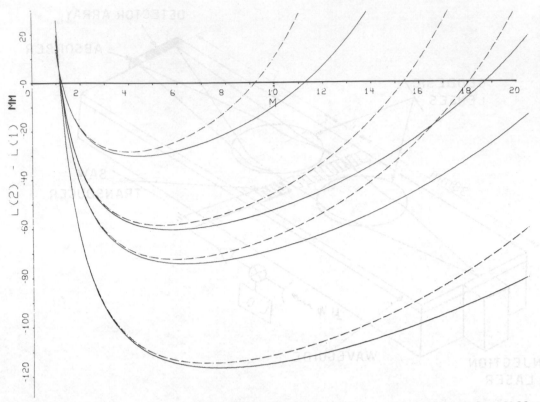

Figure 3 Size saving of the two-lens system plotted versus the magnification m, for different values of the spacing S (from the top to the bottom: 7.5, 12, 14, and 20 μm). Two values, 1.024(continuous line) and 1.41 (dashed line), of the truncation factor g are considered.

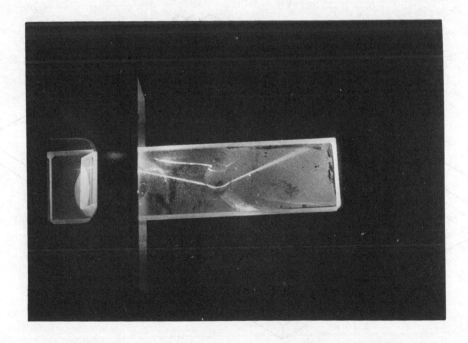

Figure 4 Two-lens geodesic system built in a BK-7 glass substrate. The guide is an epoxy resin film doped with RhB. A narrow beam deflected by the two lenses is clearly visible; the spreading of the beam after the second lens is due to the non uniform thickness of the guide.

FIBER OPTIC SENSORS

H. F. Taylor, T. G. Giallorenzi and G. H. Sigel, Jr.

Naval Research Laboratory, Washington, DC, 20375, USA

INTRODUCTION

The feasibility of using fiber optic sensors for detecting acoustic pressure, magnetic field, acceleration, temperature, and rotation rate has already been demonstrated. Efforts are now underway in a number of laboratories to develop cost-effective sensors for use in a variety of civilian and military applications. This paper reviews the state-of-the-art in sensors based on single mode fiber technology.

Sensor Components

Component requirements for single mode fiber sensors differ considerably from those of fiber optic communication systems. Whereas communications systems generally use intensity modulation of the light source, operate with low signal levels (< -30 dBm) and at high frequencies (> 1 MHz), sensors must detect small-amplitude phase modulation and usually operate at high signal levels (> -20 dBm) and at low frequencies (< 10 kHz). Single mode GaAlAs laser diodes are generally the preferred optical source for single-mode sensors. Some characteristics of these lasers which must be taken into account in sensor design include low-frequency intensity[1] and phase noise[2] and feedback-induced mode hopping[3] and spectral broadening.[4] Silicon PIN photodiodes, which combine high quantum efficiency with low cost and simple packaging, are usually used for optical detection. The 3-5 dB/km losses at .82 μm in presently available fibers is generally adequate for most sensors since fiber lengths are not extremely long (typically in the 10 m 1 km/range). It is felt that best sensor performance, particularly in the case of the gyroscope, will be obtained through the use of highly birefringent fiber to reduce polarization noise effects.[5-7] Such fiber is not yet available with low loss, but is under development at several laboratories.

Other, more specialized components have been developed for sensor application. One example is the evanescent field fiber coupler, in which single mode fibers with most of the cladding etched away are twisted together to obtain synchronous coupling between the fiber cores. The original BOTTLE coupler[8,9] in which the cores are surrounded by index-matching oil, has recently been superseded in the NRL sensor experiments by new designs which eliminate the ambient liquid. These new designs, which are more stable relative to temperature changes and have a longer operating life, make use of silicone potting,[10] glass gel potting,[10] thermal fusion[11] of the fibers. Other specialized components include fiber "squeezers" for producing phase shifts or more general alterations of the polarization state,[12] and piezoelectric cylinders wrapped with fiber,[13] also used to produce a controllable phase shift.

Sensor Configurations

The Mach-Zehnder fiber interferometer configuration being actively investigated for measuring acoustic pressure, magnetic field, and acceleration is illustrated in Fig. 1. Evanescent field fiber couplers are used as the input and output beam splitters, thus providing a continuous fiber path from the source to the photodetectors. The perturbation of interest causes a phase change in the signal arm of the interferometer, relative to that in the reference arm. In the acoustic sensor, the applied pressure causes a change in both the length and refractive index of the sensing fiber.[14-16] The length change is greatly enhanced by coating the fiber with a compliant polymer.[17-19] In the magnetic sensor, the signal fiber is coated with or bonded to a strip of magnetostrictive material.[20,21] The accelerometer uses a proof mass to stretch one arm of the interferometer.[22]

Best sensitivity in an interferometric sensor is obtained when it is operated with a relative phase shift which gives nearly equal intensities for the two outputs (i.e., near the "quadrature" condition). Because thermal drifts and other low-frequency environmental effects tend to drive the interferometer away from quadrature, a compensating phase shift is produced in the reference fiber which is wrapped around a piezoelectric cylinder. The feedback signal to drive the piezoelectric is obtained by passing the two interferometer output signals through a differential amplifier and a low-pass filter.[13] This maintains high-sensitivity operation of the interferometer for measuring the relatively high-frequency signal of interest.

Another class of fiber optic sensors, termed polarimetric sensors,[23] makes use of relative phase shifts between orthogonally polarized components in a single fiber, as illustrated in Fig. 2. In contrast to the physically separate optical paths of the Mach-Zehnder sensor, the polarimetric sensor relies upon interference between waves which travel the same path in orthogonal polarization states. Birefringent fibers are desirable in polarimetric sensors to minimize noise due to polarization mode coupling effects. Linearly polarized light with the polarization plane oriented at 45° to the birefringence axes is coupled into the fiber. Light transmitted through the fiber is split into linearly polarized components by a Rochon or Wollaston prism oriented at 45° to the fiber axes. Intensities of these two components vary with the relative phase shift of the orthogonally polarized fiber mode in a manner analogous to the variation of the two outputs of the Mach-Zehnder interferometer with relative phase shift in the fiber arms. Both acoustic and magnetic polarimetric sensors have been demonstrated. The acoustically induced anisotropic strain was obtained in one experiment by wrapping the fiber under tension around a compliant cylinder subject to ambient pressure,[24] and in another by cementing the fiber to the outside of a hollow cylinder.[25] For the magnetic sensor, the fiber was wound under tension around a magnetostrictive cylinder.[26]

The rotation-induced phase shift in counterpropagating beams in a Sagnac interferometer is the basis for the fiber optic gyroscope[27,28] illustrated in Fig. 3. The output from a laser is divided by a beam splitter into two equal-intensity components which are coupled into opposite ends of a single mode fiber. After propagating through the fiber and acquiring a relative phase shift proportional to rotation rate, the output beams interfere at the same beam splitter to produce an output signal. A beam splitter situated between the source and the input beam splitter provides a second output signal with complementary dependence of intensity on phase shift. Designs with evanescent-field fiber beam splitters[29,30] provide a continuous optical path from source to detector, thus eliminating component motion noise effects encountered when bulk beam splitters are used.

As in the Mach-Zehnder sensors, it is desirable to provide a controllable phase shift in the fiber gyroscope to maintain maximum sensitivity. However, a simple dc phase shift in the interferometer path is not effective for the rotation sensor since the interfering beams transverse the same optical path. Phase shifting techniques which have been used for the Sagnac gyro include ac phase modulation (electrooptic,[31] piezoelectric),[32] acoustooptic deflection,[33] and external cavities with dc phase shifters.[34,35] The acoustooptic phase shifter in a phase nulling scheme is particularly attractive from the standpoint of readout simplicity and dynamic range, since the modulator drive frequency, which is proportional to phase shift, can be determined by conventional digital frequency measurement circuitry.

Sensor Performance

Results representative of the state-of-the-art in fiber optic sensors for acoustic pressure, magnetic field, acceleration, and rotation rate are described below.

Results on an experiment on a fiber optic acoustic sensor immersed in a water tank have been reported.[36] The light source in that experiment was a HeNe laser, and evanescent-field fiber couplers were used to eliminate component motion noise in a Mach-Zehnder interferometer. Minimum detectable signal levels near the ambient noise for "sea state zero" (60 dB re 1 μPa at 100 Hz, 40 dB re 1 μPa at 1 kHz) were observed.

Experiments on a Mach-Zehnder magnetic field sensor have also been carried out. Data on the frequency dependence of the minimum detectable ac magnetic field for a 10 cm length of fiber bonded to a nickel tube[37] are plotted in Fig. 4. The sensitivity is close to 1 μOe per meter of fiber length over the frequency range from 100 Hz to 2 kHz. Similar sensitivities were obtained in nickel-coated fibers. Recently, much better sensitivity, in the range of 10^{-9} Oe/m, have been observed in some preliminary experiments on fibers bonded to metallic glass strips.[37]

High sensitivity has also been demonstrated in a fiber optic accelerometer.[22] Over a range of frequencies from 30 Hz to 300 Hz, the noise level in the interferometer corresponded to a signal level of less than 2 μg.

In the fiber gyroscope, recent experiments[30,38] have pushed sensitivity well below the "earth's rate" benchmark of 15°/h. In one case, a superluminescent diode was used as the light source and a polarization scrambler was employed in the Sagnac interferometer path.[39] Noise due to polarization coupling and reflection-induced instabilities in the laser was sufficiently reduced that rms noise and drift were in the 10°/hr range, although the received optical power level was only 25 nW.

CONCLUSIONS

Single mode fiber optic sensors represent a relatively new area of technology, but impressive results have already been achieved. Demonstrated performance of sensors for acoustic pressure, magnetic fields, rotation rate, and acceleration is approaching that of the best conventional sensors. In the future we anticipate that fiber optics will offer high performance and low cost in a variety of civilian and military sensor applications.

REFERENCES

1. Dandridge, A., Tveten, A.B., Miles, R.O., and Giallorenzi, T.G., 1980, "Laser Noise in Fiber Optic Interferometer Systems," Appl. Phys. Lett., 37, 526-528.

2. Dandridge, A., Tveten, A.B., Miles, R.O., Jackson, D.A., and Giallorenzi, T.G., 1981, "Single-Mode Diode Laser Phase Noise," Appl. Phys. Lett., 38, 77-78.

3. Miles, R.O., Dandridge, A., Tveten, A.B., Giallorenzi, T.G., and Taylor, H.F., 1981, "Low-frequency Noise Characteristics of Channel Substrate Planar GaAlAs Laser Diodes," Appl. Phys. Lett., 38, 848-850

4. Miles, R.O., Dandridge, A., Tveten, A.B., Taylor, H.F., and Giallorenzi, T.G., 1980, "Feedback-Induced Line Broadening in cw Channel-Substrate Planar Laser Diodes," Appl. Phys. Lett., 37, 990-992.

5. Ramaswamy, V., Kaminow, I.P., Kaiser, P., and French, W.G., 1978, "Single Polarization Optical Fiber," Appl. Phys. Lett., 33, 814-816.

6. Kaminow, I.P., Simpson, J.R., Presby, H.M., and MacChesney, J.B., 1979, "Strain Birefringence in Single-Polarization Germanosilicate Optical Fibres," Electron. Lett., 15, 677-679.

7. Dyott, R.B., Cozens, J.R., and Morris, D.G., 1979, "Preservation of Polarisation in Optical-Fibre Waveguides with Elliptical Cores," Electron. Lett., 15, 380-382.

8. Sheem, S.K., and Giallorenzi, T.G., 1979, "Single-Mode Fiber Optical Power Divider: Encapsulated Etching Technique," Opt. Lett., 29, 29-31.

9. Sheem, S.K. and Giallorenzi, T.G., 1979, "Polarization Effects on Single-Mode Optical Fiber Sensors," Appl. Phys. Lett., 35, 914-917.

10. Tran, D.C., Koo, K.P., and Sheem, S.K., 1981, "Single-Mode Fiber Directional Couplers Fabricated by Twist-Etching Techniques (Stabilization)," IEEE J. Quant. Electron., QE-17, 988-991.

11. Villarruel, C.A. and Moeller, R.P., 1981, "Fused Single Mode Fibre Access Couplers," Electron. Lett., 17, 243-244.

12. Johnson, M., 1979, "In-line Fiber Optical Polarization Transformer," Appl. Opt., 18, 1288-1289.

13. Jackson, D.A., Priest, R.P., Dandridge, A., and Tveten, A.B., 1980, "Elimination of Drift in a Single Mode Optical Fiber Interferometer Using a Piezoelectrically Stretched Coiled Fiber," Appl. Opt., 19, 2926-2929.

14. Bucaro, J.A., Dardy, H.D., and Carome, E.F., 1977, "Optical Fiber Acoustic Sensor," Appl. Opt., 16, 1761-1762.

15. Cole, J.H., Johnson, R.L., and Bhuta, P.G., 1977, "Fiber-Optic Detection of Sound," J. Acoust. Soc. Amer., 62, 1136-1138.

16. Culshaw, B., Davies, D.E.N., and Kingsley, S.A., 1977, "Acoustic Sensitivity of Optical Fiber Waveguides," Electron. Lett., 13, 760-761.

17. Bucaro, J.A., and Hickman, T.R., 1979, "Measurement of Sensitivity of Optical Fibers for Acoustic Detection," Appl. Opt., 18, 938-940.

18. Hocker, G.B., 1979, "Fiber Optic Acoustic Sensors with Increased Sensitivity by Use of Composite Structures," Opt. Lett., 4, 320-321.

19. Hughes, R., and Jarzynski, J., 1980, "Static Pressure Sensitivity Amplification in Interferometric Fiber Optic Hydrophones," Appl. Opt., 19, 98-107.

20. Yariv, A., and Winston, H.V., 1980, "Proposal for Detection of Magnetic Fields through Magnetostrictive Perturbation of Optical Fibers," Opt. Lett., 5, 87-89.

21. Dandridge, A., Tveten, A.B., Sigel, G.H., Jr., West, E.J., and Giallorenzi, T.G., 1980, "Optical Fibre Magnetic Field Sensors," Electron. Lett., 16, 408-409.

22. Tveten, A.B., Dandridge, A., Davis, C.M., and Giallorenzi, T.G., 1980, "Fibre Optic Accelerometer," Electron. Lett., 16, 854-856.

23. Rashleigh, S.C. and Ulrich, R., 1981, "Polarized Light Propagation in Single-Mode Fibers," Third International Conference on Integrated Optics and Optical Fiber Communication, Optical Society of America, San Francisco, California.

24. Rashleigh, S.C., 1980, "Acoustic Sensing with a Single Coiled Monomode Fiber," Opt. Lett., 5, 392-394.

25. Rashleigh, S.C., and Taylor, H.F., 1981, "Beamforming Fibre Optic Sensor," Electron. Lett., 17, 138-139.

26. Rashleigh, S.C., 1981, "Magnetic-field Sensing with a Single Mode Fiber," Opt. Lett., 6, 19-21.

27. Vali, V., and Shorthill, R.W., 1976, "Fiber Ring Interferometer," Appl. Opt., 15, 1099-1100.

28. Vali, V., and Shorthill, R.W., 1977, "Ring Interferometer 950 m Long," Appl. Opt., 16, 290-291.

29. Sheem, S.K., 1980, "Fiber Optic Gyroscope with [3x3] Directional Coupler," Appl. Phys. Lett., 37, 869-871.

30. Bergh, R.A., Lefevre, H.C. and Shaw, H.J., 1981, "All Single-Mode Fiber-Optic Gyroscope," Opt. Lett., 6, 198-200.

31. Bulmer, C.H., and Moeller, R.P., 1981, "Fiber Coupled Phase Shifter for Use in Optical Gyroscopes," Third International Conference on Integrated Optics and Optical Fiber Communication, Optical Society of America, San Francisco, California.

32. Bohm, K., Russer, P., Weidel, E., and Ulrich, R., 1981, "Low-Noise Fiber Optic Rotation Sensing," Opt. Lett., 6, 64-66.

33. Cahill, R.F., and Udd, E., 1979, "Phase-nulling Fiber Optic Laser Gyro," Opt. Lett., 4, 93-95.

34. Hotate, I., Yoshida, Y., Higashiguchi, M., and Niwa, N., 1980, "Rotation Detection by Optical Fibre Laser Gyro with Easily Introduced Phase-Difference Bias," Electron. Lett., 16, 941-942.

35. Goss, W.C., Goldstein, R., Nelson, M.D., Fearnchaugh, H.T., and Ramer, O.G., 1980, "Fiber-Optic Rotation Sensor Technology," Appl. Opt., 19, 852-858.

36. Donovan, J.E., Giallorenzi, T.G, Bucaro, J.A., and Simmons, V.P., 1981, "Applications of Fiber Optics in Sensors," Electro '81.

37. Sigel, G.H., Jr., and Koo, K.P., 1981, unpublished results.

38. Arditty, N., 1980, "Recent Developments in Guided Wave Optical Rotation Sensors," Topical Meeting on Integrated and Guided Wave Optics, Optical Society of America, Incline Village, Nevada.

39. Bohm, K., Marten, P., Petermann, K., Weidel, E., and Ulrich, P., 1981, "Low-Drift Fibre Gyro using a Superluminescent Diode," Electron. Lett., 17, 352-353.

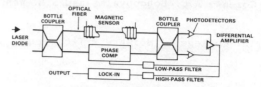

Fig. 1 Schematic diagram of an interferometer fiber optic sensor for magnetic field detection.

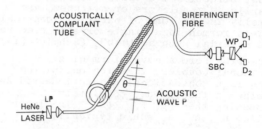

Fig. 2 Polarimetric sensor. The Wollaston prism (WP) separates the optical output into two orthogonally polarized components; the Soleil-Babinet compensator (SBC) provides a variable phase shift.

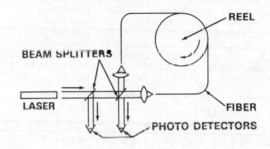

Fig. 3 Schematic diagram of fiber optic gyroscope.

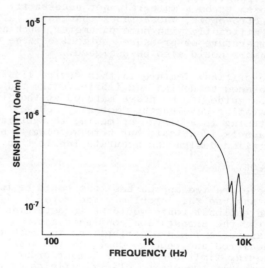

Fig. 4 Frequency dependence of minimum detectable field for fiber optic magnetic sensor.

CO-AXIAL OPTICAL COUPLERS

J.R. Cozens, A.C. Boucouvalas and N.L. Webster

Imperial College of Science & Technology, U.K.

INTRODUCTION

Optical directional couplers designed primarily as modulators or switches, are generally in the form of parallel identical channels formed in a single crystal electro-optic material such as $LiNbO_3$ (Schmidt (1)).

We describe here a novel form of a directional coupler, in which one of the guides is a conventional single mode rod and the other a tube guide, completely surrounding the rod, Fig. (1). The layers labelled n_1, n_2, n_1 would typically be constructed in glass or silica, the outer layer, n_3, being an electro-optic material.

The advantages of such an arrangement are that the device has cylindrical symmetry, and hence can be joined to fibres easily, and that propagation is predominantly in materials with low loss, so that propagation losses will be minimal. The operation of the device as a modulator is based on conventional principles. The guides are phase matched in the absence of an applied field, so that power launched into the rod is completely transferred to the tube in a beat length λ_B. With a field applied, the modified n_3 causes a phase mismatch such that power is returned to the rod in the distance λ_B. Hence the output from the rod guide can be modulated by the applied electric field.

Alternatively, the outer layer could be chosen to be a material, not necessarily electro-optic, whose index changes significantly with some parameter, such as temperature or pressure, so that a range of sensors could also be designed.

The critical feature in this device lies in the need to design and fabricate the rod and tube guides to be phase matched at the operating wavelength. Normally, phase matching is achieved by making the guiding channels identical, but here we clearly need detailed design and accurate fabrication.

THEORY

There are two approaches that could be taken to analyse this problem. The more conventional route would be to determine first the propagation constants and transverse field distributions for both the ideal rod and tube guides. In the weakly guiding limit and for the lowest order mode we assume transverse electric fields of the form

$$E_1 = \begin{cases} AI_0(\gamma_1 r) & r < a_2 \\ B_1 J_0(\kappa r) + B_2 Y_0(\kappa r) & a_2 < r < a_3 \\ CK_0(\gamma_2 r) & r > a_3 \end{cases}$$

for the tube guide, and

$$E_2 = \begin{cases} DJ_0(\kappa r) & r < a_1 \\ FK_0(\gamma_1 r) & r > a_1 \end{cases}$$

for the rod guide,

where $\kappa = (n_1^2 k_0^2 - \beta^2)^{\frac{1}{2}}$

$\gamma_1 = (\beta^2 - n_2^2 k_0^2)^{\frac{1}{2}}$

$\gamma_2 = (\beta^2 - n_3^2 k_0^2)^{\frac{1}{2}}$

Similar expression can be written for the magnetic field components.

Continuity of field components at the interfaces yields two characteristic equations, from which the corresponding β's can be separately determined. Combination of dimensions and indices can thus be found that satisfy the phase matching requirement. The coupling between two guides can be determined from the well-known coupling theory (Marcuse (2)) using the transverse field expressions for the rod and tube guides. The coupling coefficient, c, is simply related to the beat length, λ_B, by

$$c = \frac{\pi}{2\lambda_B}$$

From a knowledge of the propagation constants and the coupling coefficient and their dependence on indices and dimensions, all the essential characteristics of the device can be determined.

An alternative route has also been followed as a useful independent check of the results, in which a normal mode analysis of the combined rod and tube guide was carried out. The presence of three boundaries leads to a characteristic equation whose roots can be determined from a 12 x 12 determinant in the exact case. The beat length calculated from the difference in β of the two normal modes was in excellent agreement with the results of the coupled mode analysis.

RESULTS

The reduced propagation constant,

$$b = \frac{(\frac{\beta}{k_0})^2 - n_2^2}{n_1^2 - n_2^2}$$ of the lowest order mode of a

selection tube guides is shown in Fig. (2). Clearly, for some particular rod guide, there are an infinite set of tubes with identical β, i.e. for any inner tube radius, an appropriate wall thickness can be chosen.

The apparently arbitrary choice of inner tube radius has a strong bearing on the magnitude of the coupling coefficient, however, and hence the beat length, Fig. (3). If we select values of the indices n_1, n_2, n_3, the V value of the rod guide and the required beat length (hence separation between rod and tube), then the dimensions of the tube for phase matching are completely defined. However our choice of some of these parameters has a marked bearing on the fabrication tolerances required. In order to investigate these tolerances, we will assume that the co-axial fibre is prepared by forming a preform by a rod-in-tube method, followed by pull down. If this technique is followed, then we may expect that the ratios of the radii of rods and tubes will not change during pulling, merely their absolute values. Precision in the dimensions of layers in the fibre will therefore reflect essentially the accuracy and stability of the pulling motors. We can investigate the consequences of a pulling error in which the absolute dimensions of the fibre deviate from those required for phase matching. Deviation from phase matching is represented by the maximum power, P, transferred from the rod (initially with power P_0) to the tube, where

$$\frac{P}{P_0} = \left[1 + \left(\frac{\Delta\beta}{2c}\right)^2 \right]^{-1}$$

and $\Delta\beta = \beta_{tube} - \beta_{rod}$.

The results of some representative calculations are given in Fig. (4). Clearly the fabrication tolerance becomes more severe as the separation between rod and tube increases. For the fabrication of a passive device, then small separations, with stronger coupling and shorter beat length would be prepared. For some form of switching, however, a change in n_3 (for example) is required. We can investigate the change of beat length with n_3 for power to be switched completely between the guides for a fixed length of interaction. In Fig. (5) we see that the switching sensitivity increases with beat length. Thus, not surprisingly, we find that conditions for higher sensitivity are associated with tighter fabrication tolerances. If the inner layers of the device are fabricated in glass or some other very low loss material, the propagation losses will be small, and relatively long devices could be practicable. From Fig. (5), a device with a beat length of 5 cms would require Δn_3 of $5 \cdot 10^{-5}$ ($V_{rod} = 1 \cdot 8$) or $9 \cdot 10^{-5}$ ($V_{rod} = 2 \cdot 3$). The glasses chosen in all these examples have indices which are appropriate to match with KDP as the outer material. The switching requirements are consistent with the electro-optic coefficients of KDP in association with reasonable electric field, so that an electro-optic modulator based on the design outlined here appears feasible. The switching requirements can be relaxed some- what by taking advantage of conventional alternating $\Delta\beta$ techniques. Overall, the fabrication tolerances appear reasonable when set alongside the advantages of low loss, cylindrical geometry, and switching sensitivity.

Let us now consider the possibilities for the design of a range of sensors. If the outer layer is a polymer, it would be relatively easy to strip sections of guide in order to replace this outer cladding with another of different index. Thus, long lengths of cable could be prepared with strongly mis- matched guides, so that power launched into the central conventional rod guide, would propagate without coupling to the tube. The sensing action can now be localised to some desired region by changing the outer cladding as described.

The operation of the sensor might thus consist of the determination of the power remaining in the rod guide, after a phase matched beat length of interaction. Such measurements of absolute intensity, especially if made remote from the sensor, are less than ideal. It would be preferable to base the observation on some property of propagation that could be expected to be relatively inviolate except due to the specific interaction in the device.

One such property is the wavelength of the radiation used. While the nature of the coupler will clearly not cause significant wavelength changes to be imposed on a monochromatic source, the phase matching condition could be made very sensitive to wavelength, so that only a narrow band of wavelengths can couple. If broad band radiation is launched, a narrow band can be coupled out from the rod, being observed as an absorption band in the subsequent radiation from the rod. The wavelength of the absorption band will now be a measure of the outer cladding and hence of the sensed parameter.

The results of some calculations of coupling bandwidths and device sensitivities are given. We assume a structure as in Fig. 1, which we make phase matched at some wave- length λ_0. If we include material dispersion for typical glass (n_1, Chance Pilkington BSC517642, $n_2 = n_3$, BSC510044) we find the half-width of the coupled band, (Fig. 6) to be a fairly sensitive function of the separation between the rod and the tube, Fig. 7. Such a fibre will have bandwidths ranging from 400 Å with a separation of 4·5 μm to 2Å with separation of 9·5 μm. The corresponding beat lengths range from 6 mm to 1 m (Fig. 7). To reduce the band- width further we could aim to increase the material dispersion of the outer cladding. This could probably be achieved by doping the material so that an absorption band is introduced near to the range of operating wavelengths. With this approach there is the additional advantage that the absorption in the outer cladding would tend to remove the coupled power, without regard to the length of interaction. We show in Fig. 7 that bandwidths as low as $3 \cdot 10^{-2}$ Å could be achieved with a cladding material whose dispersion is assumed to be an order of magnitude larger than that of dispersive glass, and with a separation between rod and tube of 9·5 μm. Again larger separations imply larger beat lengths, which would have a bearing on spatial resolution.

The bandwidth calculations can be readily translated into a measure of minimum detectable change in n_3. The dependence of the phase matched wavelength, λ_0, on n_3 for a particular guide is shown in Fig. 8. This result is not dependent on the dispersive characteristics of the outer cladding.

We may expect that the minimum detectable change in n_3, Δn_{3min}, occurs for a shift in λ_0 of a bandwidth. From Fig. 8 $dn_3/d\lambda_0$ has a constant value ($-5.26 \cdot 10^{-5} \mathring{A}^{-1}$) so that

$$\Delta n_{3min} = \left(\frac{dn_3}{d\lambda_0}\right)\Delta\lambda_{B/W}$$

We may thus expect Δn_{3min} to range from about 10^{-5} for $\Delta\lambda_{B/W}$ of $2\mathring{A}$ to 10^{-7} for $\Delta\lambda_{B/W}$ of $10^{-2}\mathring{A}$, Fig. 9.

From the temperature dependence of refractive index, we can estimate minimum detectable temperature changes. A value of dn/dT for a typical glass is $3 \cdot 10^{-6}$ giving ΔT_{min} between 3^0K and $3 \cdot 10^{-2}$ ^0K. Alternative cladding materials could be chosen with higher values of dn/dT.

The co-axial coupler thus appears to offer some promise as the basis of a range of sensors. The advantages include cylindrical symmetry, low loss, high sensitivity and the ability to localise sensors onto a 'sensing highway' without the need to form joints. The problems include the launching of broad band radiation into a single mode guide, and the development of a range of cladding materials, with defined indices and high dispersion characteristics.

This work is supported in part by a grant from the Aileen S. Andrew Foundation.

REFERENCES

1. Schmidt, R.V., and Alferness, R.C., 1979, Directional Coupler Switches, Modulators, and Filters using Alternating $\Delta\beta$ Techniques, IEE Transactions on Circuits and Systems, CAS-26, 1099-1108.

2. Marcuse, D. Light Transmission Optics, van Nostrand Reinhold, 1972, 407-431.

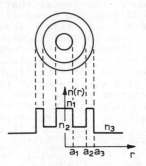

Fig. 1 Index distribution in the Co-axial Coupler

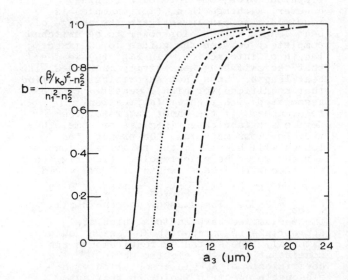

Fig. 2 Reduced propagation constants of the low order mode of a tube guide, as a function of outer radius, a_3 for various values of the inner radius a_2.

$$n_1 = 1 \cdot 51508$$
$$n_2 = 1 \cdot 50800$$

a_2

———————— 4 μm
.... 6 μm
— — — — — 8 μm
— · — · — · — 10 μm

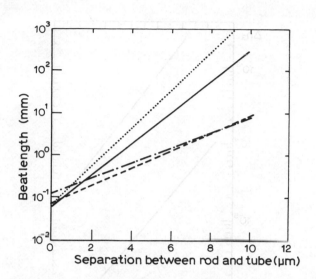

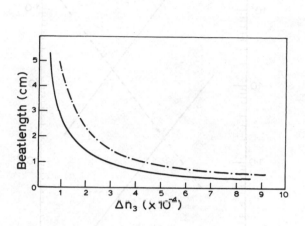

Fig. 3 Beatlength, λ_B vs separation between rod and tube (a_1-a_2), for various rod V values, and wavelengths

	V_{rod}	λ
··············	2.3	0.63 µm
——————	1.8	0.63 µm
— — — —	1.8	1.00 µm
—·—·—·—	1.8	1.30 µm

Fig. 5 Change in n_3 required for complete switching in one beatlength, vs beatlength, at $\lambda = 0\cdot63$ µm.

—·—·—·—	$V_{rod} = 2\cdot3$
——————	$V_{rod} = 1\cdot8$

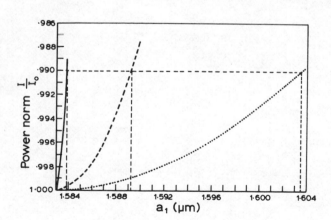

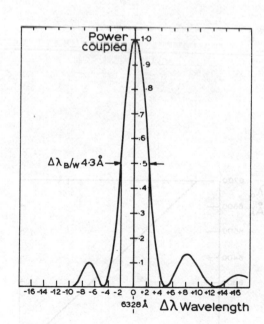

Fig. 4 Maximum power transfer in a phase matched beatlength vs radius a_1, ratios of dimensions held constant. Phase matched dimensions (µm):

	a_1	a_2	a_3	
——————	1.583	8.033	9.606	$V_{rad} = 2.3$
— — — —	1.583	6.000	7.569	
·············	1.583	4.000	5.558	$\lambda = 0.63$ µm

Fig. 6 Normalised coupled power vs wavelength (illustrative example).

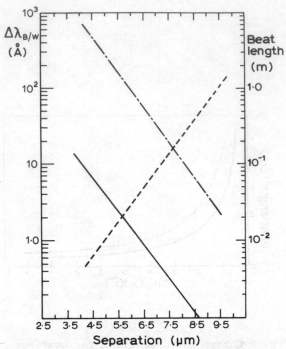

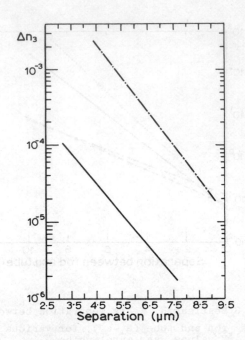

Fig. 7 Bandwidth, $\Delta\lambda_{B/W}$, vs separation,
(a₁-a₂) outer cladding,

—·—·— n₃, to have index and dispersion
of Chance Pilkington B.S.C. 510644.

———— n₃ to have dispersion x 10 of
dispersive glass.

— — — — Beat length, shown vs separation.

Fig. 9 Minimum detectable change in n₃,
Δn₃, vs separation

—·—·— outercladding Chance Pilkington
B.S.C. 510644.

———— outercladding with dispersion x 10 of
dispersive glass.

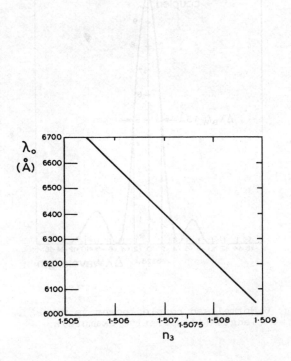

Fig. 8 Phase matched wavelength λ_0 vs outer
cladding index n₃.

SINGLE MODE INTERFEROMETER CURRENT SENSORS WITH OPTICAL FIBRES AND DIODE LASERS

A. Dandridge, A. B. Tveten and T. G. Giallorenzi

Naval Research Laboratory, Washington, DC 20375

INTRODUCTION

Recently there has been considerable interest in using optical fibres as sensing elements in sensor systems. In these systems, usually the optic fibre's coating or composition is optimized to respond to the field to be detected. Such systems have been constructed from both multimode and single mode fibres. For single mode fibre systems, a number of different configurations of both one fibre and two fibre interferometers have been used. Using the Faraday effect, Rashleigh and Ulrich[1] developed a current sensor in the range 0.2 → 2000 A. Recently it has been shown that a system employing multimode fibre can be constructed to measure a.c. currents, the current range investigated was between 5 to 2000 mA.[2] In this paper, we report on a.c. current measurements, using different sensor configurations, in single mode fibre Mach Zehnder interferometers.[3] The current range investigated was between ∿ 1 A and ∿ 10^{-9} A. The possibility of using these systems to detect currents in the range 10^6 to 10^{-12} A is also discussed.

CONFIGURATIONS

Two different approaches were used to transform the current to a phase shift detectable in the interferometer system; a) The I^2R heating produced by the current heated the optical fibre. b) The magnetic field produced by the current acting on magnetostrictive material bonded to the fibre.

The configuration of the two types of current sensor are shown in Fig. 1. Figure 1a depicts a fibre bonded to a piece of thin-walled (0.1 mm) nickel tubing. Thin walls were used to minimize eddy current effects and reduce demagnetization effects. Sensor lengths were in the order of 10 cm. The sensor is placed within the coil through which the current to be detected is passed, typically the coil resistance was 5 Ω (for measurements of current in the μA range). Figure 1b depicts a single mode fibre coated with a 2 μm layer of Al, electrical contacts were made at both ends of the coated region (typically 10 cm long and having a 3 Ω resistance). The current to be detected is passed directly through the metallic coating of the fibre using the contacts shown.

PHASE MEASUREMENT

The measurement of the small phase shifts produced by the heating and magnetostrictive effects was accomplished by the use of an all-fibre interferometer capable of detecting phase shifts as small as 10^{-6} radians.[3] The system differed from earlier fibre interferometers,[3,4,5] in that the 'bottle' coupler (beam splitter) was of a ruggedized form,[6] and the usual HeNe source replaced by a Hitachi HLP 1400 single mode GaAlAs laser operated at 830 nm. These improvements on the earlier interferometer systems are leading towards small ruggedized interferometer systems. The interferometer system was maintained at maximum sensitivity (quadrature) by an electronic compensation scheme. The compensation scheme resembled that described by Jackson et al.[5], but was run in the high-gain bandwidth product mode, thus allowing retrieval of all signals in a 1.0 Hz - 10 kHz band.

RESULTS

Shown in Fig. 2 is the response of the magnetostrictive sensor to applied current, the response is linear over 4 1/2 decades. The sensor was 10 cm long and the frequency of the a.c. current was 10 kHz. Shown in Fig. 3 is the response of the magnetostrictive current sensor with frequency, the results are plotted in terms of radians per Ampere and have been normalized to a 1 m length of the sensor. At 10 kHz the measured sensitivity of the device was ∿ 7×10^{-9} A per m of sensor. By using longer sensor lengths, more finely wound coils and optimizing the magnetostrictive jacketing of the fibre sensitivities in the 10^{-12} A may be possible. To detect larger currents (> 1 A) the coil could be constructed with a low resistance and weak coupling to the magnetostrictive coating.

The frequency response of the heating current sensor is shown in Fig. 4. The sensitivity varies approximately linearly as the reciprocal of the frequency. At a frequency of 10 Hz the sensitivity was measured to be 5×10^{-8} Amp/m. Larger currents could be measured by using an appropriate shunt in parallel with the fibre conductor. Improvements in the design and response of the described systems are presently being investigated along with other configurations to provide the phase change in the interferometer. One such configuration presently under study is the strain produced by the interaction between the current along the conductor coated fibre and an external magnetic field.

CONCLUSION

Two types of fibre optic current sensors have been constructed and shown to provide a sensitive technique for a.c. current measurement. Experimental sensitivities of ∿ 7×10^{-9} A per m have been achieved.

REFERENCES

1. S.C. Rashleigh and R. Ulrich, Appl. Phys. Lett. 34, 768, 1979 "Magneto-coupling current sensing with birefringent fibers."

2. G.L. Tangon, D.I. Persechini, R.J. Morrison and J.A. Wysocki, Private communications.

3. D.A. Jackson, A. Dandridge, S.K. Sheem, "Measurement of Small Phase Shifts Using A Single-mode Optical Fibre Interferometer," Opt. Lett. 5, 139, 1980.

108

4. A. Dandridge, A.B. Tveten, G.H. Sigel, Jr.,
   E.J. West, T.G. Giallorenzi, "Optical
   Fiber Magnetic Field Sensors," Elect. Lett.,
   16, 408, 1980.

5. D.A. Jackson, R. Priest, A. Dandridge,
   A.B. Tveten, "Elimination of drift in a
   single mode optical fiber interferometer
   using a piezoelectrically stretch coiled
   fiber," Appl. Opt. 19, 2926, 1980.

6. C. Villarruel, S.K. Sheem, K.P. Koo, Naval
   Research Laboratory, Washington, DC,
   unpublished design.

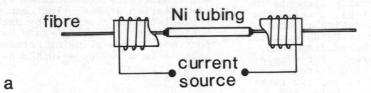

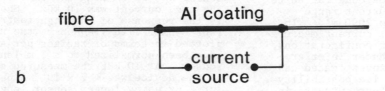

Figure 1
Schematic diagram of fibre-optic current sensor,
a) magnetostrictive sensor, b) heating sensor.

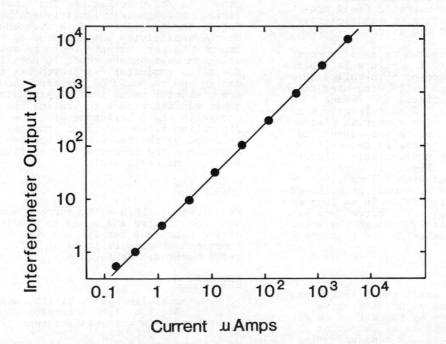

Figure 2
Interferometer output against applied a.c.
current for the magnetostrictive sensor operating at 10 kHz and having a length of 10 cm.

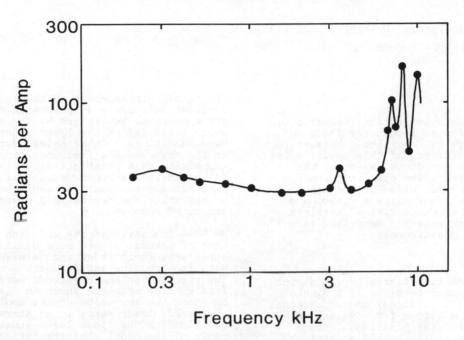

Figure 3

Sensitivity of the magnetostrictive current
sensor in radians/Amp as a function of frequency.

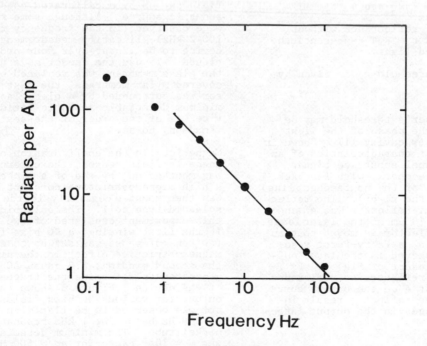

Figure 4

Sensitivity of the heating current sensor in
radians/Amp as a function of frequency.

SEMICONDUCTOR LASER SENSOR: SOURCES AND CONFIGURATIONS

A. Dandridge, R.O. Miles, A.B. Tveten and T.G. Giallorenzi

Naval Research Laboratory, Washington, DC 20375, US

INTRODUCTION

In this paper, a compact optical sensor system consisting of a semiconductor laser and an external cavity is described. Measurements of the sensitivity of this device using a number of different diode lasers as the source are reported. The different configurations of the device, allowing the sensor to respond to acoustic waves (as a microphone and hydrophone), magnetic fields, accelerations and a.c. currents are described as well as the resultant measurements.

MODE OF OPERATION

In Fig. 1, a schematic diagram of the device is shown. The sensor consists of a single mode diode laser ($\lambda \sim 830$ nm) and an external reflector placed within a few wavelengths of the laser's facet. The position of the external reflector is perturbed slightly by the incident field to be measured. The output of the laser is measured by a large area photodiode placed at the rear facet of the laser. The phase of light fed back from the external reflector is determined by the distance d and the perturbation of the external reflector. The laser facet reflectance R in the presence of an external reflector r is determined by

$$R = R_o + 2(1 - R_o) \sqrt{rR_o} \cos \theta$$

where R_o is the facet reflectance without feedback (~ 0.32) and $r \ll R, \theta$ represents the phase of the reflected light.

The gain at lasing threshold g_o is given by,

$$g_o = -\frac{1}{2L} \ln RR_o$$

consequently, the laser's threshold may be altered by changing the phase of the light fed back into the laser cavity (1). Shown in Figure 1 is the output characteristics of an Hitachi HLP 1400 channel substrate planar diode laser with no feedback, with feedback in phase (to the left of the no feedback line) and out of phase (to the right). The reflectivity of the external reflector was 4% and it was placed within 10 μm of the laser facet. When the external reflector is moved continuously in one direction, a Fabry-Perot resonance response is observed in the laser output as shown in the insert of Fig. 1. By placing the external reflector at a distance corresponding to point A on the output curve, small movements of the reflector result in relatively large changes in the output intensity of the laser.

SOURCES

The minimum detectable field, i.e. sensitivity, of such a device is determined by i, the magnitude of the movement of the external reflector to the incident field (e.g. magnetic or acoustic) and ii, the minimum detectable displacement of the reflector. The smallest detectable displacement of the sensor appears to be limited by the amplitude noise level (2) of the laser under the feedback conditions. Other noise sources, such as the observed low frequency line jitter (appearing as a phase noise) of these lasers, (3) or ambient acousto-mechanical noise were respectively too small or could be eliminated by operating the device in a vibration free evacuated chamber.

We have investigated the emission characteristics of several single mode GaAlAs laser structures (both index and gain guided) under both free running and feedback conditions. The results for the amplitude noise in the free running case are shown in Fig. 2, as can be seen, the low value of the amplitude noise of the TJS laser makes it an attractive proposition for the diode laser sensor. Measurements of the modal characteristics of the lasers, under the feedback conditions of the sensor, will also be reported.

CONFIGURATIONS OF THE SENSOR

Acoustic. To use the sensor as a microphone, a thin glass membrane was used as an external reflector (i.e. 4% reflectance). The frequency response of the sensor was determined by measuring its output, with an acoustic field set up by a calibrated constant output acoustic source. Although some resonances were observed (in the frequency range of 100-2 kHz), linearity measurements showed the device to be linear over four orders of magnitude. To use the sensor as a hydrophone, the glass membrane was replaced by a thin gold covered mylar membrane, the rest of the sensor was surrounded by a plexiglass shell. The minimum detectable acoustic field of the device was of the order of the sea state zero (Knudson) noise.

Magnetic. In the configuration of an a.c. magnetic field sensor, the external reflector was mounted on the end of a nickel tube with a high magnetostrictive constant. The tube was then mounted coaxially inside of a solenoid-type coil. The solenoid coil for this experiment consisted of two windings. In the first winding, a DC bias field of a few tens of gauss was set up to enhance the magnetostrictive effect in the nickel. In the second winding, a 4 gauss AC field was generated to determine the frequency response of the device. Figure 3 shows the detector output for various DC bias fields. The output was observed to be flat with frequency up to 200 Hz below the 1 kHz resonance of the nickel tube. The minimum detectable field was $\sim 4.10^{-5}$ gauss for an ~ 600 Hz signal, the sensitivity decreasing with decreasing frequency, owing to the frequency dependence of the amplitude noise (3).

AC Current Sensor. This was achieved by using a similar configuration to that outlined in

111

the magnetic sensor. Different ranges of sensitivity were obtained by using different
solenoid coils, and shorter nickel tubes.

Acceleration sensor. Both cantilever and
membrane devices have been constructed and
measurements showed the device's response to
be linear with the applied field over four
orders of magnitude. However, low frequency
resonances restrict the frequency range over
which the device can usefully operate. Work
is continuing to eliminate, or raise the
frequency of the observed resonances.

REFERENCES

1. Kressel, H., and Butler, J. K., Semiconductor lasers and heterojunction LEDS,
   (Academic Press, 1977) pp. 261-265.

2. Dandridge, A., Tveten, A.B., Miles, R.O.,
   Giallorenzi, T.G., Single mode diode laser
   phase noise, Appl. Phys. Lett., 38, 77,
   (1981).

3. Dandridge, A., Tveten, A.B., Miles, R.O.,
   Laser noise in fiber optic interferometer systems, Appl. Phys. Lett., 37, 526,
   (1980).

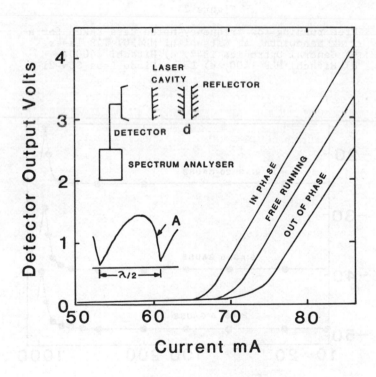

Figure 1

Laser Sensor Characteristics

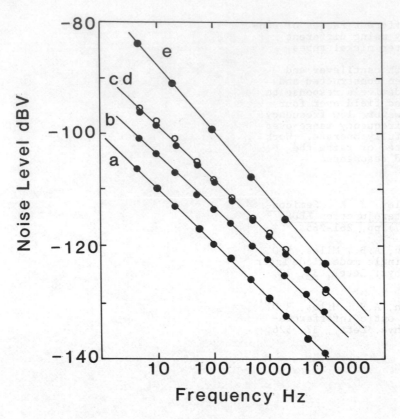

Figure 2

Free running low frequency noise data taken for a
1 Hz bandwidth, a) Mitsubishi ML4307 TJS laser,
b) General Optronics TB47, c) Hitachi 2400U,
d) Hitachi HLP 1400, e) Laser Diode Labs SCW-21

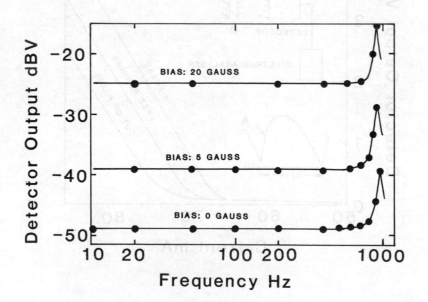

Figure 3

Magnetic field sensor output as a function of frequency
for different DC bias fields